Recent Researches in Polymer and Plastic Processing

Recent Researches in Polymer and Plastic Processing

Guest Editor

Joanna Izdebska-Podsiadły

MDPI

Basel • Beijing • Wuhan • Barcelona • Belgrade • Novi Sad • Cluj • Manchester

Guest Editor
Joanna Izdebska-Podsiadły
Department of Printing
Technology
Warsaw University of
Technology
Warsaw
Poland

Editorial Office
MDPI AG
Grosspeteranlage 5
4052 Basel, Switzerland

This is a reprint of the Special Issue, published open access by the journal *Materials* (ISSN 1996-1944), freely accessible at: https://www.mdpi.com/journal/materials/special_issues/284ANJ6B15.

For citation purposes, cite each article independently as indicated on the article page online and as indicated below:

Lastname, A.A.; Lastname, B.B. Article Title. *Journal Name* **Year**, *Volume Number*, Page Range.

ISBN 978-3-7258-5997-9 (Hbk)
ISBN 978-3-7258-5998-6 (PDF)
https://doi.org/10.3390/books978-3-7258-5998-6

Cover image courtesy of Joanna Izdebska-Podsiadły

Contents

About the Editor

Joanna Izdebska-Podsiadły

Joanna Izdebska-Podsiadły, professor at the Warsaw University of Technology, is a specialist in the field of printing technologies, materials engineering, and polymer surface modification.

She graduated with honors from the Warsaw University of Technology in 2004 and obtained a doctorate in technical sciences in the field of mechanics and machine construction in 2011. In 2022, she received her postdoctoral degree, presenting a series of publications on the modification of the surface layer of biodegradable films and printing ink formulations to improve wettability and printability.

Her scientific achievements include over 30 scientific articles, 12 chapters in monographs, numerous conference publications, and educational books. She is the co-editor of a monograph published by Elsevier entitled *Printing on Polymers: Fundamentals and Applications* and the editor of a monograph entitled *Polymers for 3D Printing: Methods, Properties, and Characteristics*.

Her research focuses on issues related to printing on biodegradable materials, the surface activation of polymers via physical methods (plasma, corona discharge), the modification of flexographic inks, and the use of environmentally friendly additives to improve printing properties. Her research is interdisciplinary and application-oriented, and its results are used in the packaging industry.

As a teacher, she conducts classes in chemistry, packaging technology, printing plant design, and computer graphics. She has supervised and mentored dozens of theses. She actively participates in the popularization of science, cooperates with industry, and is involved in the development of international academic cooperation.

Preface

In the face of rapid technological advancement and growing environmental challenges, polymer science has emerged as a vital interdisciplinary field. This Reprint entitled "Recent Research in Polymer and Plastic Processing" was conceived to bring together and showcase the latest developments in polymer engineering, functionalization, and sustainable applications. The guiding theme of this volume, Polymers Reimagined—From Surface Engineering to Sustainable Recycling, reflects both the breadth of topics covered and the innovative spirit driving contemporary research.

This collection of ten peer-reviewed articles published in the journal *Materials* spans a wide range of subjects, including plasma-based surface modification to improve coating adhesion, the printability of biodegradable starch-based films, the optimization of injection molding processes using artificial intelligence, and the mechanical performance of lightweight 3D-printed PLA structures. It also explores novel approaches to reinforcing polymer composites, the use of industrial by-products in fiber production, and the development of hydrogels for internal curing in cementitious materials. Finally, this volume addresses the degradation and recycling of widely used polymers such as PET, EVA, and PVDF, with a strong emphasis on circular economy principles.

This Reprint will be a useful resource for researchers, engineers, doctoral candidates, and students in the technical and natural sciences who seek up-to-date knowledge and inspiration for further scientific and industrial work. We hope that the insights presented here will contribute to the advancement of innovative technologies and foster a more responsible approach to designing the materials of the future.

Joanna Izdebska-Podsiadły
Guest Editor

materials

MDPI

Review

Surface Modification of Polymers by Plasma Treatment for Appropriate Adhesion of Coatings

Gregor Primc and Miran Mozetič *

Department of Surface Engineering, Jozef Stefan Institute, Jamova cesta 39, 1000 Ljubljana, Slovenia; gregor.primc@ijs.si
* Correspondence: miran.mozetic@guest.arnes.si

Abstract: In this study, recent advances in tailoring the surface properties of polymers for the optimization of the adhesion of various coatings by non-equilibrium gaseous plasma are reviewed, and important findings are stressed. Different authors have used various experimental setups and reported results that scatter significantly and are sometimes contradictory. The correlations between the processing parameters and the adhesion are drawn, and discrepancies are explained. Many authors have explained improved adhesion with the adjustment of the surface free energy or wettability of the polymer substrate and the surface tension of liquids used for the deposition of thin films. The adhesion force between the polymer substrate and the coating does not always follow the evolution of the surface wettability, which is explained by several effects, including the aging effects due to the hydrophobic recovery and the formation of an interlayer rich in loosely bonded low molecular weight fragments.

Keywords: polymers; gaseous plasma; wettability; adhesion; coatings

1. Introduction

The adhesion of various coatings on polymer substrates has attracted significant attention from the scientific community due to the need to invent reliable, inexpensive, and ecologically benign methods for optimizing the adhesion force and durability of products made from adhered materials. The interaction between the substrate and the coating has also been studied, and numerous mechanisms have been proposed [1]. Briefly, the mechanisms could be categorized as mechanical coupling and chemical interaction. The mechanical coupling will be enhanced for substrates of appropriate roughness, while the chemical interaction will be optimized if the deposited material forms covalent or other strong bonds with the functional groups on the polymer surface. A crucial property that affects both mechanical and chemical interaction is the ability of a coating to fill any gaps or pores on the surface of materials to be joined. The effect is illustrated in Figure 1.

Figure 1a illustrates the case of optimal adhesion. The materials stick well because all gaps or other morphological features are filled. The optimal adhesion is rarely achieved since the coating does not wet the substrate thoroughly, i.e., does not fill tiny pores or gaps. The inadequate adhesion is illustrated in Figure 1b. In this case, the interface between the two materials is minimal because the coating does not wet the substrate. This effect is often observed when the substrate is hydrophobic and rough on the sub-micrometer scale, and the coating is hydrophilic. The effect illustrated in Figure 1b could be suppressed by pressing the substrate and the coating together to force the filling of at least some gaps or pores. An alternative solution is the application of ionic liquids for coating various substances onto the polymer surfaces [2]. Yet another alternative is the hydrophilization of the substrate. Hydrophilization can be achieved by chemical methods, often by depositing an intermediate layer that balances the substrate's surface free energy and the coating's surface tension. These chemicals are usually various primers, which may form a very thin film on the substrate. An alternative to chemical methods for balancing the substrate's surface properties and/or coating is the application of gaseous plasma. Gaseous plasma is a source of particles with considerable kinetic or potential energy and radiation, of which the photon energy often exceeds the binding energy of atoms in solid materials. Numerous authors have reported beneficial results when using gaseous plasma. Selected recent scientific articles are briefly reviewed, and the results are interpreted in the view of plasma science.

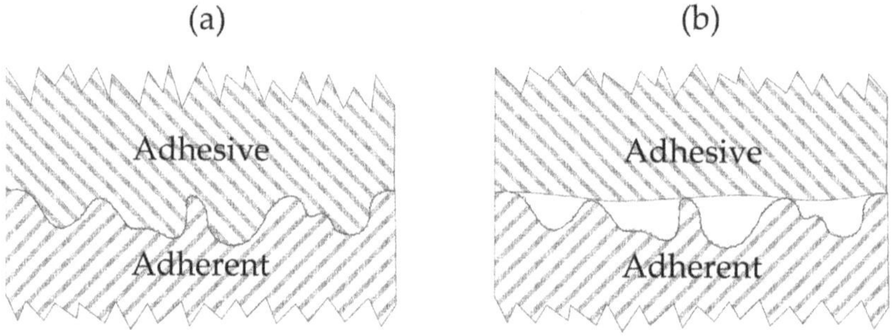

Figure 1. Influence of morphology on the optimal (**a**) and inadequate (**b**) adhesion between two materials.

2. Gaseous Plasma

Gaseous plasma is often called the fourth state of matter, but it is actually gas with a significant concentration of free electrons and positively charged ions. Polymer materials are usually treated with non-equilibrium gaseous plasma, a state of the gas in which the temperature of different particles deviates significantly. The translational temperature of heavy particles (all plasma particles except electrons) is usually just above the room temperature, while the electron temperature is often of the order of 10,000 K. The rotational temperature of molecules is often close to the translational temperature, but the vibrational temperature could be several 1000 K. The degree of ionization and dissociation is much larger than calculated from the Boltzmann distribution, taking into account the translational temperature.

The electron temperature is usually much larger than the translational temperature of other gaseous particles, like molecules and ions [3]. The translational temperature of molecules in plasmas suitable for the treatment of polymer materials is often close to room temperature, and the electron temperature usually exceeds 10,000 K. Such a large electron temperature enables inelastic collisions with gaseous molecules, which leads to the excitation, dissociation, or ionization of the molecules [4]. The degrees of excitation, dissociation, and ionization in non-equilibrium gaseous plasma are much larger than predicted by the Boltzmann equation, taking into account the translational temperature of molecules. Some plasmas exhibit vibrational temperatures of several 1000 K, almost irrespective of the gas translational temperature [5]. The electronically excited molecules and atoms may be de-excited by radiation, usually in the vacuum ultraviolet range, i.e., with the photon energy between 6 and 12 eV [6]. Therefore, the major factors governing the interaction of non-equilibrium gaseous plasma with polymer materials will be the fluxes of plasma species on the polymer surface: positively charged ions, neutral radicals, including atoms in the ground state, molecules and radicals in metastable excited states, and photons in the UV or VUV range, and not the gas translational temperature or the temperature of the polymer surface. The kinetic energy of ions impinging on the polymer surface is also essential if the polymer surface is biased against the plasma.

The interaction of plasma species (charged particles, molecular radicals, and excited species) with solid surfaces is always exothermic. It often leads to the modification of the surface composition and structure, so it is suitable for tailoring surface properties of various materials [7]. Non-equilibrium plasmas are often used on an industrial scale to modify polymer wettability [8]. The exact mechanisms of plasma–polymer interaction are still inadequately understood, and recent theoretical articles have enlightened the phenomenon's complexity [9,10]. In general, the plasma treatment causes at least one (more frequently several) of the following effects:

- Heating;
- Substitution of surface atoms with atoms from plasma (often called functionalization);
- Breaking polymer chains, depolymerization, and formation of polymer fragments;
- Etching;
- Modification of polymer morphology;
- Modification of the elasticity of the polymer surface film;
- Breaking bonds in the surface film and possible cross-linking;
- Modification of surface wettability.

The latter is particularly interesting when studying the adhesion of coatings on polymer materials. The wettability reflects the ability of the coating to penetrate gaps and pores (Figure 1). Suppose the surface is highly wettable (hydrophilic). In that case, the soft or melted materials, let alone liquids, will enter the pores because of the capillary forces, so the effect illustrated in Figure 1a is feasible. If the wettability is poor, the effect illustrated in Figure 1b will likely occur.

3. Review of Recent Literature

Numerous authors have reported using plasma treatment to increase the adhesion of various coatings on polymer materials. Only selected recent papers published since 2022 are here reviewed, and the observations are explained and illustrated.

3.1. High-Pressure Plasmas

A large frequency of collisions between gaseous particles governs the atmospheric pressure plasmas. The foundations of atmospheric pressure plasmas have been presented by Bruggeman et al. [11]. Briefly, the collision frequency (average number of collisions per unit of time) increases linearly with the increasing density of gaseous particles, which is proportional to the gas pressure and inversely proportional to the translational temperature of colliding particles. The collision frequency between molecules, atoms, and ions is roughly 1 MHz, and the electron-neutral collision frequency is as large as about 1 THz [11]. Such a large frequency of collisions causes intensive kinetic energy exchange between the colliding particles, so the electron temperature in atmospheric pressure non-equilibrium plasma is rarely above 1 eV. The kinetic energy of positively charged ions impinging on the surface of a plasma-treated material is often negligible, i.e., below 1 eV. Furthermore, the probability of three-body collisions increases as the square of the density of colliding species [12]. The loss of molecular radicals in the gas phase occurs predominantly by three-body collisions, so the lifetime of radicals such as atoms is as low as about 1 µs at atmospheric pressure. Thus, the molecule dissociation fraction is much lower at atmospheric pressure than in low-pressure plasmas, and a homogeneous plasma in a large volume is very difficult to achieve.

Li et al. [13] treated polytetrafluoroethylene (PTFE) with helium plasma and added ammonia water to the processing gas. The optimal pressure in the processing chamber was 0.5 MPa, and the total gas flow rate was 4 L/min. The wettability was evaluated by measuring the static water contact angle (WCA), while the adhesion of the epoxy glue was evaluated by peel strength experiments. The plasma generator operated at the frequency of 60 kHz, with a maximal voltage of 5.12 kV. The ammonia water and helium mixing ratio varied between 0 and 2.5 vol.% of ammonia water. The plasma-treated samples were also characterized by X-ray photoelectron spectroscopy (XPS), and roughness was evaluated by atomic force microscopy (AFM). The PTFE treatment with pure helium plasma improved wettability because the WCA dropped from 110° to 40° after 15 s and 30° after about 120 s. After that, the WCA remained constant. Although the authors did not mention it, the increased wettability is attributed to the absorption of vacuum ultraviolet radiation, which causes structural changes in the PTFE surface film [14]. The addition of ammonia water caused a further decrease in the WCA to about 20°, and the optimal admixture was about 1 vol%. XPS results revealed the formation of oxygen-containing functional groups even on the sample treated with pure helium plasma, and the concentration of these groups increased significantly with a small admixture of ammonia water to the plasma gas. The authors even recognized some C–N bonds in de-convoluted high-resolution C1s spectra. AFM did not reveal noticeable morphological changes, but the pristine samples were relatively rough, so any changes might have been hidden. Namely, OH and O radicals (from water) always cause laterally non-homogeneous etching of the PTFE materials, as explained in [15]. Plasma-treated PTFE samples were glued with epoxy adhesive, and the peel strength was measured. The adhesion was found to be enhanced significantly. Even for samples treated with pure He plasma, the peel strength increased by a factor of 5.1, and with the addition of 1 vol.% of ammonia gas by a factor of 6.6. Interestingly, the authors attributed increased adhesion to the formation of amine groups despite the marginal concentration of C–N groups compared to O–C=O determined from the XPS high-resolution C1s spectra. A feasible explanation of the observations reported by Li et al. [13] is illustrated in Figure 2. Helium plasma sustained at high pressure (Li et al. [13] reported 5 bar) is a significant source of VUV radiation (Figure 2a), which arises from the resonant relaxation of the He_2^* excimers [16]. The absorption depth of such low-wavelength radiation is of the order of 10 nm [17], so the photons cause radiation damage and, thus, the formation of dangling bonds (Figure 2b). Dangling bonds interact chemically with oxygen upon exposure to air to form polar oxygen-containing functional groups (Figure 2c). The addition of ammonia water (or any other molecules) suppresses the radiation from the He_2^* excimers [16]. Still, it causes the dissociation and excitation of radicals (Figure 2d), so the PTFE surface is simultaneously treated with VUV radiation arising from H, H_2, O, and the radicals (including NH_x) which interact chemically with the dangling bonds and cause extensive surface functionalization with polar functional groups (Figure 2e).

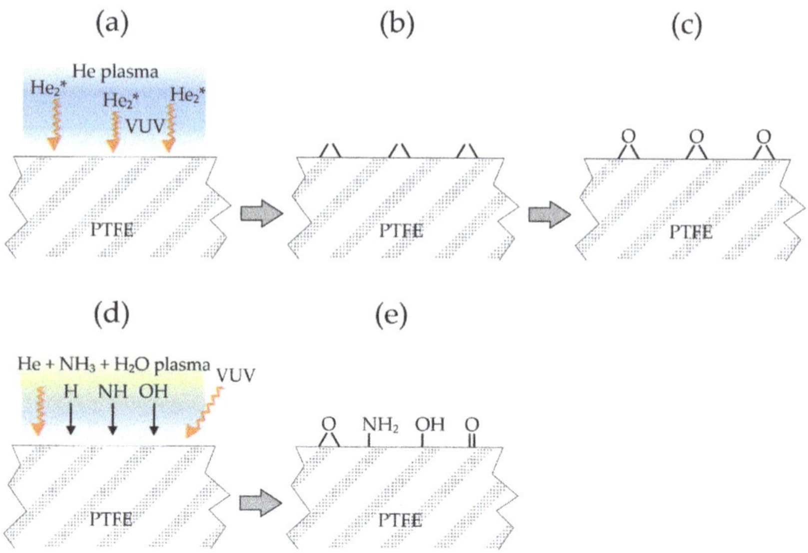

Figure 2. Illustration of the interaction between plasma sustained in pure helium (**a–c**) or a mixture of helium with ammonia water (**d**,**e**) with the surface of the PTFE.

Antipova et al. [18] probed another fluorine-rich polymer widely used in some industries and for tissue engineering, polyvinylidene fluoride (PVDF). They used helium plasma at atmospheric pressure and at a discharge power of 9 W for the treatment of the

PVDF samples in order to increase their wettability. Unlike Li et al. [13], Antipova et al. found a significant increase in surface roughness after the plasma treatment. AFM imaging taken on the surface area of 2.5 μm × 2.5 μm revealed the increased roughness from about 2.5 to 15 nm after treating the samples for 90 s. Intriguing, the WCA dropped from 82° to only 59° and 53° after 60 s and 90 s treatments. They characterized the samples by X-ray diffraction and concluded that plasma treatment does not induce any structural modifications, as the positions and relative intensities of the diffraction peaks remained identical for untreated and plasma-treated samples. Antipova et al., however, found a significant modification in Young's modulus. While it was about 450 MPa for untreated samples, it dropped to 70 and 40 MPa for samples treated for 60 and 90 s, respectively. The interaction of helium plasma with PVDF is similar to that illustrated in Figure 2a–c, except that the relatively low discharge power did not assure extensive bond breakage by absorption of VUV radiation. The authors attributed such a significant reduction in the substrate surface elasticity to chemical etching induced by the He plasma, which disrupted the cross-linking in the polymer matrix. The authors used plasma-treated samples to study the adhesion and viability of stem cells. They found a significant increase in the density of adhered cells in samples treated for 90 s, while the differences between untreated samples and samples treated for 60 s were marginal. On the other hand, cell viability increased by 10 and 12% for samples treated for 60 and 90 s, respectively. The main improvement was observed qualitatively by examining cells with an optical microscope. While cells were overstressed when grown on untreated PVDF, they were adopted well on plasma-treated substrates and formed aggregates.

Plasma polymer treatment has helped increase the tensile shear strength between aluminum and polymer sheets after hot pressing at 320 °C. Takenaka et al. [19] used atmospheric pressure plasma sustained in argon with an RF generator that operated at the frequency of 60 MHz. The plasma plum extended outside of the quartz tube, so admixing of the effluent air most likely occurred. Polyether ether ketone (PEEK) was selected as the polymer, and the aluminum strip was made from an A1050 grade alloy (containing at least 99.5% Al). The tensile shear strength for untreated materials was about 7 MPa. When the aluminum strip was treated with plasma and the polymer was untreated, the shear strength was as low as 1 MPa, so the treatment of the Al materials with plasma worsened their adhesion capability with the polymer upon hot pressing. When both aluminum and PEEK strips were treated, the shear strength increased to about 9.5 MPa. The best results, however, were reported when only the polymer strip was treated by plasma because the shear strength, in this case, was as large as 13 MPa. This value is higher than when the authors used an adhesive to join both materials. Cohesive and interfacial delamination were reported when the PEEK was treated by plasma and aluminum was untreated. The beneficial results were reported only for the prolonged plasma treatment of PEEK because the shear strength stabilized only after 5 min of plasma treatment. Before, the shear stress was pretty linear with the plasma treatment time. The results Takenaka et al. [17] reported are illustrated in Figure 3. The selected discharge frequency causes significant heating of the gaseous plasma at high pressure. Oxygen from the effluent air dissociates in the jet of Ar plasma (Figure 3a), and the O atoms cause the formation of a relatively thick and rough oxide film on the aluminum surface (Figure 3b). The authors showed scanning electron microscope (SEM) micrographs and EDX and XPS results, which firmly confirmed the formation of the rough oxide film on the aluminum surface. The adhesion of a PEEK foil on such an oxidized aluminum surface is illustrated in Figure 1b. When only the PEEK sheet was treated by plasma (Figure 3c), the adhesion with an Al strip was optimal because the oxygen atoms formed in plasma caused the formation of polar functional groups on the polymer surface (Figure 3d). After hot pressing, some oxygen from the polymer surface interacted with the aluminum strip to ensure highly improved adhesion. When both materials were treated with plasma, the effect of polar groups was marginal since the aluminum had already been oxidized by plasma treatment.

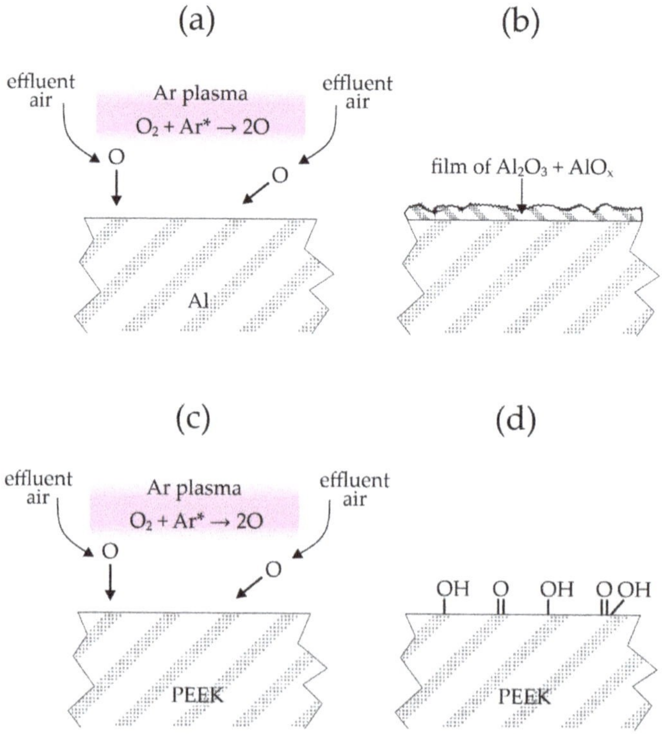

Figure 3. Interaction of 60 MHz atmospheric pressure argon plasma with aluminum (**a**,**b**) and PEEK (**c**,**d**).

Kosmachev et al. [20] studied the influence of plasma treatment on the interlayer sheer strength of laminates consisting of carbon fiber tapes and interlayers of polyphenylene sulfide (PPS), which served as a thermoplastic binder. The surface of the carbon fiber tapes was treated with pulsed discharges sustained in the air. The maximum voltage of the power supply was 56 kV, and the pulse duration of 10 ns was selected. The treatment times were varied between 5 and 20 min. XPS revealed 23 at.% oxygen even for untreated carbon fibers, and it increased to 33 at.% after 15 min of plasma treatment. High-resolution XPS C1s peaks revealed almost three times larger concentrations of the carbonyl groups after plasma treatment, but the concentration of C–O bonds remained relatively intact. The laminates were prepared in a mold heated to 340 °C at a pressure of 6.5 MPa. The interlaminar shear stress was measured systematically versus the plasma treatment time of the carbon fibers, and the ultimate values were 37, 41, 47, 50, and 52 MPa for untreated samples and samples treated for 5, 10, 15, and 20 min, respectively. The relative increase in the shear stress at the optimal conditions was 40%. SEM micrographs of the cleaved surfaces revealed that the failure occurred strictly along the binder–fiber interface for untreated samples, whereas for those treated with plasma for 15 min, some adhered fragments of the fractured binder were evident on the surface of the carbon fibers.

The same team also probed the adhesion between PEEK polymers and carbon fiber tapes. In the paper [21], the authors studied the mechanical properties of laminates using 0.25 mm thick PEEK foils as thermoplastic binders. The discharge parameters were the same as in [20], and so were the selected treatment times. SEM micrographs showed the localized etching of carbon fibers after prolonged plasma treatments. The temperature during hot pressing was 400 °C, the pressure was 6.5 MPa, and the pressing time was 30 min. The ultimate shear strength increased with increasing plasma treatment time and assumed 48, 52, 68, 73, and 72 MPa for untreated samples and samples treated for 5, 10, 15, and 20 min, respectively. The uniaxial tensile tests (in the direction of reinforcement) were also reported in [21]. Interestingly, the uniaxial tensile strength for samples treated with plasma for 15 min was 20% lower than for laminates synthesized without plasma treatment of the carbon fibers. The team [21] did not probe plasma treatment of PEEK, which was found beneficial by Takenaka et al. [19].

The effect of plasma treatment in the configuration adopted by Kosmachev et al. is illustrated in Figure 4. Ambient air contains significant concentrations of water vapor and carbon dioxide. All molecules are partially dissociated, preferentially those with the lowest dissociation energy (O_2, H_2O, CO_2), and nitrogen is excited to metastable electronic states and also vibrational states (Figure 4a). The excited nitrogen molecules are likely to form nitric oxides. The nitric oxides and radicals (predominantly O, OH, and H) interact chemically with the carbon fibers and form polar surface functional groups (Figure 4b).

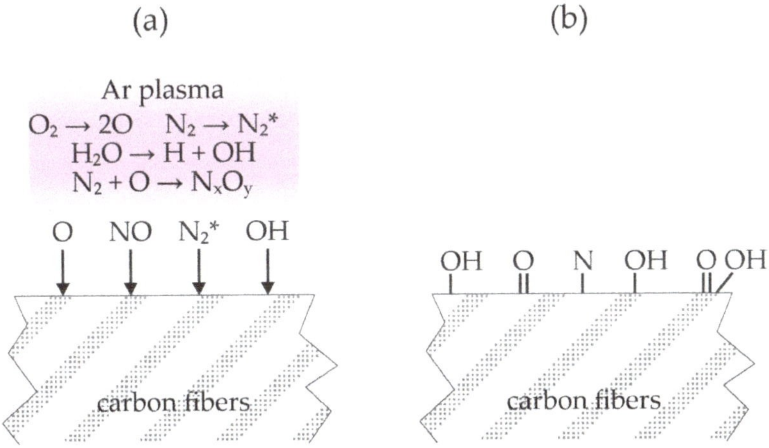

Figure 4. Interaction of air plasma with carbon fibers (**a**) and resulting surface finish (**b**).

Jung et al. [22] treated a carbon fiber-reinforced thermoplastic polymer with atmospheric pressure plasma sustained in nitrogen by a dielectric barrier discharge operating at the frequency of 13.56 MHz and voltage of 13.5 kV. The treatment time was 2 min. No wettability measurements were performed, but high-resolution XPS C1s spectra revealed a significant increase in the concentration of the hydroxyl and carboxyl bonds. Intriguingly, no nitrogen bond was observed after treatment with a plasma sustained in nitrogen. The polymer substrate was then coated with an epoxy-based adhesive, covered with an aluminum strip, pressed, and cured. The lap shear stress was about 10 MPa for untreated polymer substrates and 22 MPa for plasma-treated ones. An even larger shear stress of 27 MPa was observed when the plasma-treated carbon fiber-reinforced thermoplastic polymer was covered with a monolayer of mercapto silane. The improved adhesion was explained by the chemical interaction between the adhesive and oxygen-containing functional groups on the polymer surface. The large concentration of oxygen and no nitrogen on the polymer surface after the plasma treatment indicates that the effluent ambient air caused quenching of the nitrogen molecular metastables, similar to the experimental setup used by Kosmachev et al. [20], who also found a minimal concentration of nitrogen on the polymer surface after treatment with air plasma, so the treatment adopted by Jung et al. [22] is illustrated in Figure 4.

Kim et al. [23] treated polypropylene (PP) strips with air plasma sustained at atmospheric pressure and a discharge power as large as 1000 W. The distance between the plasma nozzle and the polypropylene samples was 1 cm, and the transverse speed was 0.25 m/s. The WCA of the untreated samples was about 87° and dropped to about 61°

after the plasma treatment. Plasma treatment did not cause a significant modification of the PP morphology, and XPS showed an increase in the oxygen concentration from 7 to 12 at.%. The high-energy tail of the high-resolution XPS C1s spectra showed significant oxidation of the polypropylene sample even before plasma treatment. The authors found the cleavage of C–C bonds in the PP by radicals generated during plasma treatment. The artificial leather made from polyurethane (PU) fabrics was pressed onto the PP samples using PU-based water adhesive as a glue. The sample temperature during pressing at the pressure of 5 bar was 65 °C. There was no adhesion between the PP substrate and the artificial leather before the plasma treatment. After plasma treatment, however, the force needed to separate the two materials with a width of 25 mm was as large as 50 N. The adhesion strength was about the same as in the parallel test using commercial primers for pre-treatment of the PP substrates. The nitrogen concentration on the PP surface after the plasma treatment was just above the detection limit of the XPS instrument, i.e., 0.9 at.%. The surface effects upon plasma treatment, as adopted by Kim et al. [23], are illustrated in Figure 4.

Some authors reported only a marginal increase in the adhesion after plasma treatment of the substrates. Ondiek et al. [24] studied the influence of plasma treatment on the adhesion properties of wood–polypropylene composites. The wood fibers were pre-coated with PP, and the fibers, PP granules, and a compatibilizer (maleic anhydride graft polypropylene) were compounded in a kneading device, pulverized, and finally molded at 200 °C in order to obtain specimens for mechanical tests according to the JIS K7139-A32 standard [25]. The composites were treated with gaseous plasma. No details about the plasma or discharge parameters were disclosed, but the treatment reduced the WCA from 108 to 25°. The composite was, therefore, highly hydrophilic, taking into account some other reports about the hydrophilization kinetics of polypropylene [23]. Different concentrations of wood fibers were added to the composite, and acrylic resin was used as a binding material. The ultimate tensile strength increased from 35 to 35.8 MPa for neat polypropylene, 40.5 to 41.1 MPa for the composite with 25% wood fibers, and 50.7 to 52.4 MPa for the composite with 25% wood fibers.

3.2. *Low-Pressure Plasmas*

Low-pressure plasmas have been used for treating solid materials on the industrial scale for decades [7]. The main advantage of low-pressure plasmas over atmospheric pressure is a much lower collision frequency and a negligible frequency for three-body collisions, which is as low as 1 Hz at the pressure of 100 Pa. Indeed, the loss of molecular radicals in the gas phase is marginal, so the dissociation fraction of molecules is large even at low discharge power per plasma volume. The properties of low-pressure plasmas are governed by surface and not gas-phase reactions. The short mean free path at low pressure enables a moderate electron temperature and high kinetic energy of ions impinging on the surface. The kinetic energy of ions bombarding the surface of a treated material can be adjusted by biasing the samples in a broad range up to a few keV. Due to the marginal loss of reactive plasma species in the gas phase, the low-pressure plasmas expand in large volumes, so the uniform plasma is sustained in chambers with a volume of several m^3 [26].

Piskarev et al. [27] treated polyethylene naphthalate foils with a low-pressure plasma sustained in the air by a direct current (DC) glow discharge—the discharge operated at 800 V and a current of 50 mA. The foils were placed either on the anode or the cathode. A few seconds of plasma treatment caused increased wettability, and the water contact angle stabilized at about 10° after several 10 s of plasma treatment. Polymer foils treated for 60 s were glued with a solution of polyurethane rubber in acetone and ethyl acetate (PU) or an ethylene vinyl acetate copolymer (EVA). They were heated at 100 °C, and then the peel resistance was determined according to the ASTM 1876-2001 standard [28]. The peel resistance was poor for untreated polyethylene naphthalate foils at 40 and 110 N/m for EVA and PU, respectively. The plasma-treated foils exhibited a much larger peel resistance of about 900 N/m for samples placed on the anode of the DC glow discharge. A significant difference in the adhesion force for samples placed on the cathode was reported.

Namely, the force was about 750 N/m for EVA and 400 N/m for PU. The authors found the only difference between the samples treated on the anode and cathode in the roughness evaluated from the measured AFM images, which was 2.1 and 2.5 for samples on the anode and cathode, respectively. Still, another difference is the heating of the electrodes. The DC glow discharge is characterized by the large voltage drop in the cathode sheath between the bulk plasma and the cathode surface [29]. Positively charged ions move randomly in the bulk plasma but are accelerated within the sheath, where they bombard the cathode with a high kinetic energy corresponding to the discharge voltage, so heating is unavoidable. The positive ions suffer some collisions within the sheath since the mean free path of gaseous particles at 15 Pa is roughly 1 mm. Positive ions thus bombard the polymer edge with moderate kinetic energy, but the rest of the polymer foil is at floating potential, so the kinetic energy of ions bombarding the polymer is only about 10 eV, as illustrated in Figure 5a. On the other side, the anode is not heated significantly due to the minimal voltage drop across the anode sheath. The dissociation fraction of oxygen molecules in the air plasma sustained at 15 Pa is considerable (Figure 5b), which explains the quick functionalization of the polymer surface with polar oxygen functional groups and, thus, appropriate wettability, as illustrated in Figure 5c.

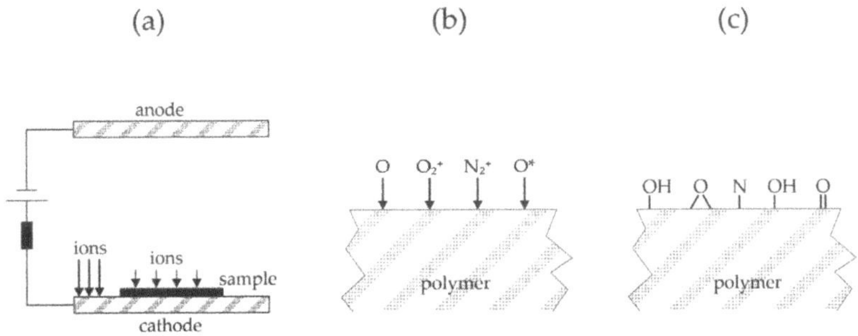

Figure 5. Illustration of the low-pressure DC glow discharge (**a**); radicals and ions from the air plasma sustained by the DC glow discharge (**b**) interact with the polymer surface and cause rapid functionalization (**c**).

Liu et al. [30] treated 5 µm carbon fibers with a low-pressure inductively coupled radio frequency (RF) plasma in the E-mode sustained in the air at the pressure of 30 Pa. The discharge power varied between 100 and 400 W. The high-resolution XPS C1 spectra revealed that the composition of the carbon fibers was similar to that of the polyamides. The high-resolution C1s spectra of the plasma-treated samples at 100 W caused the appearance of O–C=O groups at a concentration as large as 11%. Further treatment caused a decrease in the intensity of the O–C=O group to 10, 9, and 2% for discharge powers of 200, 300, and 400 W, respectively. No apparent correlation between the concentration of other groups and the discharge power was evident from the C1s spectra. The interlaminar shear strength of composites prepared from the fibers and polyimide resin was 61 MPa for untreated samples and increased to 64 and 68 MPa for 100 and 200 W discharge powers, respectively. However, the shear strength decreased with a further increase of the discharge power and assumed 63 and 62 MPa at 300 and 400 W, respectively. The decrease in the shear strength at elevated powers could be explained either by heating the fibers during plasma treatment and thus fast hydrophobic recovery or the formation of loosely bonded low molecular weight fragments at higher powers. Specifically, AFM showed a significant increase in the roughness: 110 nm for untreated samples and 130, 210, and 330 nm for samples treated at 100, 200, and 400 W, respectively. A peculiarity of electrodeless discharges sustained in glass tubes is a considerable dissociation fraction of molecules, particularly oxygen [31]. As mentioned earlier, the surface reactions are always exothermic, so samples with small dimensions like fibers of diameter 5 µm quickly heat to elevated temperatures. Highly polar functional groups are unstable at elevated temperatures [32], so the etching of polymer materials prevails over functionalization at large discharge powers. Adequate surface functionalization and adhesion of a coating is thus achieved only at a relatively small discharge power. Liu et al. [30] chose a treatment time as long as 900 s, so the fibers of small diameter heated to prohibitively high temperatures, especially at large discharge powers. The thermal effects explain the low oxygen concentration on the fiber surface and the loss of adhesion forces at large power. The effect is illustrated in Figure 6.

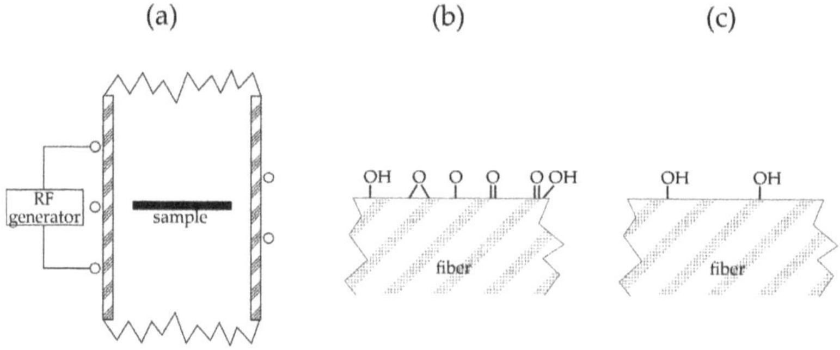

Figure 6. Illustration of inductively coupled plasma (**a**), the surface finish at low discharge power (**b**), and at high discharge power (**c**).

Wang et al. [33] studied the adhesion of copper layers on polyimide and PEEK resins and proposed an illustration of the bonds responsible for better adhesion on plasma-treated polymers. Various pre-treatments by chemical methods and sandblasting were used before the plasma treatment. The interlayer between the polymer substrate and the copper film was a thin titanium layer deposited by magnetron sputtering. The polymers were first treated with argon plasma at the pressure of 0.2 Pa and biased against the plasma potential. The treatment times were 10, 30, 80, and 150 s. After activating the polymer surfaces with argon plasma, the samples were coated first with titanium and then copper using sputter deposition without breaking vacuum conditions. No apparent correlation between the roughness on the nanometer scale and tensile strength was found for polyimide substrates, but the authors reported an excellent correlation for PEEK—a larger surface roughness resulted in a higher tensile strength, as illustrated in Figure 1a. Many polymer samples were treated with Ar plasma at the fixed bias of −100 V. For the polyimide substrates, the tensile strength at −100 V did not depend on plasma treatment time since it was always around 1 MPa. On the other hand, PEEK samples exhibited tensile strengths of 4, 7, 15, and 6 MPa for samples treated for 10, 30, 80, and 150 s, respectively. The thermal effect explains the drop in the tensile strength for the longest treatment. As mentioned earlier, biasing causes the acceleration of positively charged ions, bombarding the polymer surface with significant kinetic energy. The authors selected pressure as low as 0.2 Pa, where the mean free path is well below 1 cm, so the sheath next to the polymer sample is collisionless—at biasing to −100 V, the kinetic energy of positively charged ions bombarding the polymer surface is 100 eV. Sandblasting before plasma treatment caused a significant increase in the tensile strength for polyimide samples at 4.2, 5.8, 5.7, and 6.2 MPa for samples treated for 10, 30, 80, and 150 s, respectively. On the contrary, the sandblasting of PEEK substrates before plasma treatment caused an insignificant increase in the tensile strength; it was 8, 10, 8, and 11 MPa for samples treated for 10, 30, 80, and 150 s, respectively. The influence of the bias voltage was examined as well. The tensile strength decreased monotonously with the absolute value of the bias voltage for polyimide, while for PEEK, it peaked at the bias voltage of −100 V. The experimental system adopted by Wang et al. [33] is illustrated in Figure 7a. The residual atmosphere in hermetically tight low-pressure systems always consists of water vapor, which fully dissociates in Ar plasma, so the polymer sample is subjected to neutral and positively charged OH, O, and H radicals (Figure 7b). The velocity of the charged particles is perpendicular to the polymer surface, while neutral radicals move randomly in the gas phase. The radicals cause surface functionalization (Figure 7c) and improved wettability, but prolonged treatment time results in a high polymer temperature, which explains the loss of adhesion at the longest treatment time.

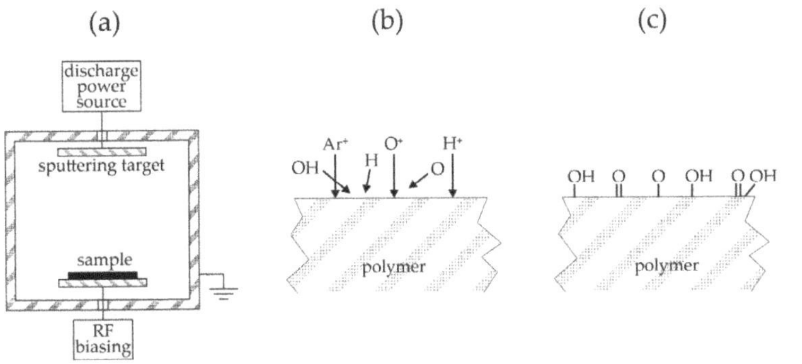

Figure 7. The schematic of the device useful for polymer activation and deposition of a thin metal film (**a**), interaction of plasma species with polymer surfaces (**b**), and resulting surface finish at moderate treatment times (**c**).

Wu et al. [34] used a low-pressure plasma sustained in the air by a capacitively coupled RF discharge operating at the frequency of about 50 kHz and a nominal power of 200 W to modify the surface of PP films. Plasma treatment caused a gradual increase in the wettability with the water contact angles of 107, 67, 64, 62, and 8° after plasma treatment for 0, 15, 30, 60, and 120 s, respectively. No explanation for a dramatic decrease in the WCA between 60 and 120 s of plasma treatment was provided, but it could be explained by the degradation of the polymer surface film due to the bombardment with moderately energetic plasma ions. A relatively thick film of polyurethane adhesive of 5 mm was deposited on the PP substrates, and the lap shear stress was measured according to the ASTM D3136 standard [35]. The shear stress increased monotonously with plasma treatment time and correlated well with wettability. The maximal load was 35, 64, 75, 80, and 110 N/cm^2 for samples treated for 0, 15, 30, 60, and 120 s, respectively. The improved adhesion was stable because, after a week of storage, the maximal load remained practically the same as for freshly synthesized samples. Furthermore, the annealing of samples treated with plasma for 30 s for one day caused a further increased shear strength to 120 N/cm^2. The authors explained the improved adhesion by chemical interaction between the hydroxyl groups formed on the PP surface during plasma treatment and the isocyanate group from the adhesive to form a urethane bond. The experimental system used by Wu et al. [34] is illustrated in Figure 8a. Low-pressure capacitively coupled discharges are renowned for DC self-biasing of the electrodes [36], and the smaller electrode (powered electrode in Figure 8a) assumes larger biasing. A polymer sample is placed on the grounded electrode. The powered electrode is subjected to extensive bombardment with positively charged ions, while the larger electrode (usually grounded, as shown in Figure 8a) is subjected to moderately energetic ions. As long as the electrically non-conductive samples like polymer foils are thin compared to the sheath thickness, the surface of the treated material assumes the same DC biasing as the backing electrode. The synergy of bombardment by positively charged ions and absorption of VUV radiation modifies the properties of the PP surface film, as illustrated in Figure 8b. The prolonged treatment causes further surface film modification and nano-structuring, as illustrated in Figure 8c. A theoretically spotless description of the modification of the polymer surface film by ion treatment is provided in [37]. The combination of rich surface morphology and functionalization with polar functional groups, as illustrated in Figure 8c, leads to the super-hydrophilic surface finish [38], which explains the WCA as low as 8° at the longest treatment time.

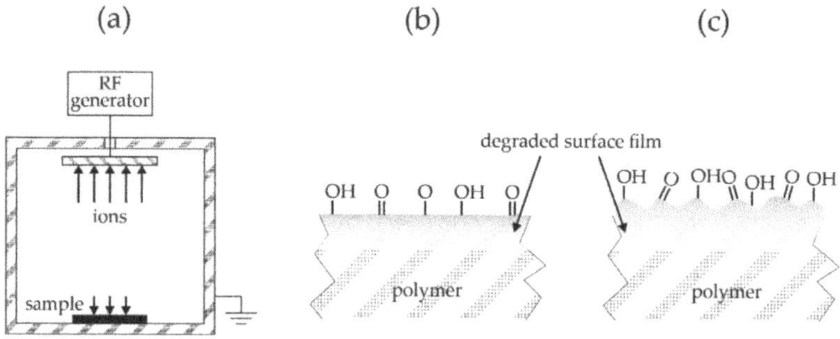

Figure 8. Illustration of capacitively coupled discharge for treatment of polymers with plasma (**a**) and the surface finish after a short treatment of several 10 s (**b**) and treatment for 120 s (**c**).

Wu et al. [39] also treated PP films with oxygen plasma for 30 s and studied the adhesion of 2 K polyurethane (PU) adhesion strength. The lap shear adhesion test was performed according to the ASTM D3163 standard method [35]. No details about oxygen plasma were provided, but they likely used the same configuration illustrated in Figure 8a. The adhesion strength increased from 24 for untreated PP films to 29 N/cm^2 for those treated with oxygen plasma. The authors added amino silane and epoxy silane to the adhesive and reported marginal differences in the adhesion strength.

4. Correlations and Paradoxes

The authors cited in Section 3 used different experimental systems and have not reported plasma parameters (i.e., the fluxes and doses of plasma species, including the positively charged ions, neutral reactive species, and energetic photons), so the reported results seem incomparable. Still, some general correlations could be drawn based on results reported by different authors for different polymer materials. Most authors reported the plasma treatment time. Figure 9 represents a plot of the water droplet contact angle (WCA) versus the treatment time. All probed polymers were moderately hydrophobic, with the WCA for untreated samples between 80 and 110°, and all became hydrophilic after the plasma treatment. The decrease in the WCA with increasing treatment time varies for different polymers, but the trend is evident: longer treatment times enable better wettability. The WCA remains moderate in some cases, but some authors reported an almost super-hydrophilic surface finish (WCA below 10°). As mentioned earlier and explained in detail elsewhere [38], the super-hydrophilic surface finish of polymers is only achieved if the surface is rough on the sub-micrometer scale and functionalized with polar groups, as illustrated in Figure 8c.

Longer plasma treatment times do not necessarily ensure better adhesion. Figure 10 is a plot of the adhesion increase versus the treatment time. The adhesion increase is defined as the shear strength after the plasma treatment divided by the shear strength of untreated samples. The reported results are scattered and are not sound with the evolution of WCA (Figure 9). Namely, the higher wettability should enable better interaction of the joined materials because it suppresses the effect illustrated in Figure 1b. The paradox is explained by loosely bonded molecular fragments formed during prolonged treatment with gaseous plasma. As mentioned above, plasma treatment always causes breaking bonds between atoms in the polymer's surface film. All authors who probed the surface composition by XPS reported a significant increase in the oxygen concentration in the surface film probed by this technique, whose probing depth is several nm. Oxygen may replace hydrogen to form hydroxyl or epoxy bonds, which is supposed to be the first effect of the plasma treatment [10,40,41], but they also break bonds in the polymer chain [42] and trigger the degradation of polymer material [43]. The degradation leads to etching, i.e., the release of volatile molecules, but incomplete oxidation leads to the formation of molecular fragments that remain on the polymer surface [44]. Large agglomerates of such fragments were reported for plasma-treated PEEK and polyether sulfone [45,46]. These molecular fragments are loosely bonded to the polymer surface and can be removed even by gentle rinsing with water [47], although the surface of these fragments is well oxidized. Thus, the well oxidized fragments of the original polymer substrate contribute to improved wettability; they are weakly bonded to the surface to the extent that they are easily removed during adhesion tests. The formation of the molecular fragments is likely to result from the synergy of VUV radiation, positively charged ions, and perhaps also neutral radicals, and is yet to be elaborated. The evolution of the molecular fragments is illustrated in Figure 11, and their influence on the peeling test is illustrated in Figure 12. The appearance of the fragments was not reported in the articles analyzed in Section 3 of this paper, so it is impossible to quantify their influence on the adhesion properties.

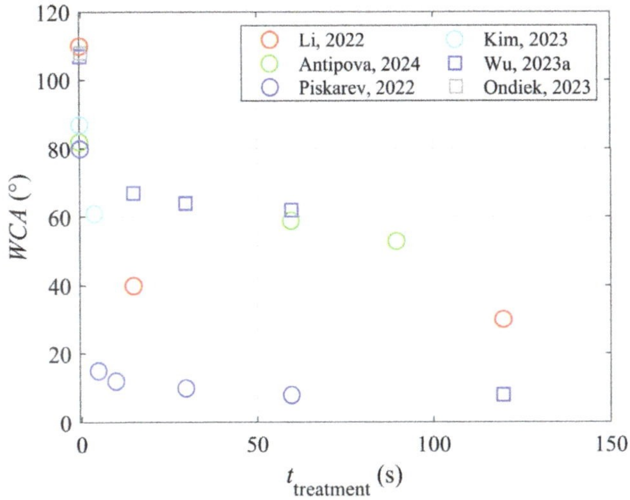

Figure 9. Water droplet contact angle versus the plasma treatment time [13,18,23,24,27,34].

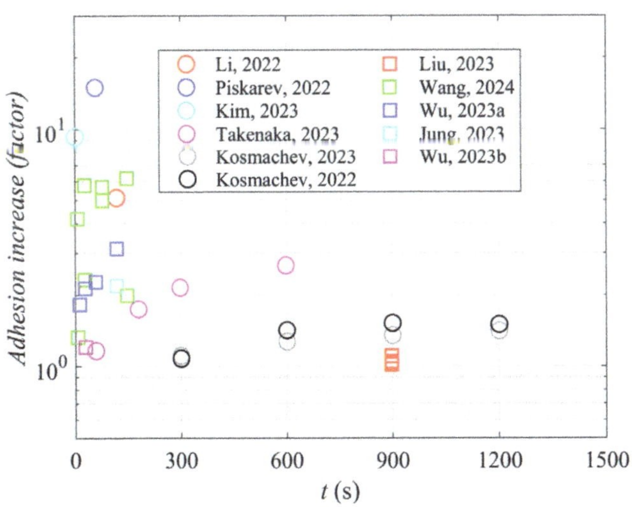

Figure 10. The ratio between adhesion after and before the plasma treatment (adhesion increase) versus the treatment time [13,19–23,27,30,33,34,39].

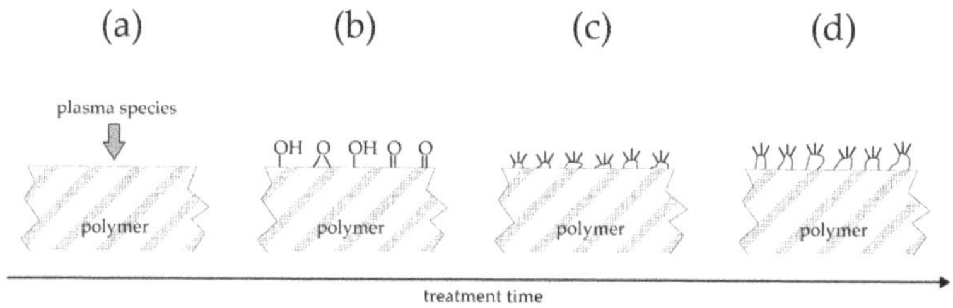

Figure 11. The formation of agglomerates of molecular fragments on the polymer surface. (**a**) The polymer is exposed to plasma species, which cause functionalization (**b**). Moderate doses of plasma species (prolonged treatment time and/or rather large discharge power) cause the formation of well functionalized molecular fragments (**c**), which may agglomerate after large doses of plasma species (**d**).

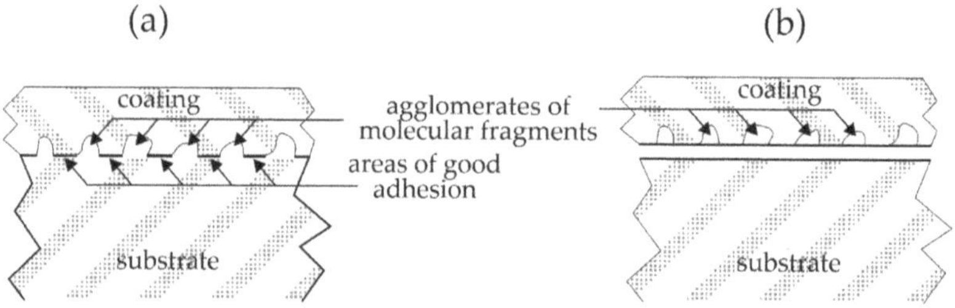

Figure 12. Illustration of the effect of molecular fragment agglomerates on the adhesion. (**a**) The coating wets the substrates well, so the coating fills all gaps on the surface, but the area of good adhesion is limited to the surface free from agglomerates. (**b**) The agglomerates are peeled from the polymer surface after the adhesion test, so the shear strength is inadequate.

5. Recommendations and Conclusions

In this study, recent papers on the adhesion of plasma-treated polymers were reviewed. Some authors reported the details about the experimental setup used to treat polymer material, but most only mentioned the type of discharge used to sustain the gaseous plasma, the gas pressure, the treatment time, and the increase in adhesion. Such an incomplete description of the experimental conditions does not enable the reproducibility of treatments at different laboratories. The authors are recommended to provide details about the experimental setups because the plasma parameters depend significantly on the dimensions and geometry of the plasma reactor, the coupling of the power supply, the biasing of the polymers against the plasma potential, and so on.

Even if the details mentioned above are provided, the plasma parameters are still challenging to deduce, so plasma characterization is recommended. A simple introduction to diagnostic methods for plasmas useful for polymer treatments is provided in [15]. Non-equilibrium gaseous plasma sustained in reactive or noble gases with an admixture of reactive gas is rich in neutral radicals, such as neutral atoms in the ground state. The atom density and, thus, the flux of atoms onto the polymer surface, depends enormously on the peculiarities of the discharges used for sustaining non-equilibrium gaseous plasma [31]. However, the typical dissociation fraction is often around 10% in plasmas with the neutral gas translational temperature close to the room temperature [7] as long as the plasma is sustained in a chamber made from dielectrics. The dissociation fraction is significantly smaller in metallic chambers [48]. The appropriate neutral radicals for tailoring the surface wettability of polymers are O atoms and OH radicals [43]. Small doses cause functionalization of the polymer surface without significant etching and degradation of the polymer chains, and the OH radicals favor the formation of hydroxyl groups [49]. The first effect of treatment with O atoms is also the formation of hydroxyl groups [10,41], but the doses exceeding 10^{20} m^{-3} cause the formation of other oxygen-containing functional groups when treating polymers both at atmospheric pressure [50] and in low-pressure reactors [51].

Other reactive species should be considered when interpreting the results of plasma treatment. Plasma is a source of VUV radiation, with the photon energy exceeding the binding energy of atoms in polymer materials. The effect of this radiation has been studied for decades and was also found significant for treating polymers on an industrial scale [52]. As much energy absorbed by gaseous plasma as about 10% was found to be transferred to VUV radiation [53], so it is of crucial importance for the hydrophilization of fluorine-containing polymers [13].

The effect of positively charged ions on the surface wettability of polymers is regarded as marginal when treating the samples with weakly ionized non-equilibrium gaseous plasma, because the ionization fraction is orders of magnitude smaller than the dissociation fraction [7]. Still, if the polymer surface is biased, as explained in [33], the ions play an important role, since their high kinetic energy causes radiation damage to the surface film. Large doses of energetic, positively charged ions will cause excessive heating of the polymers since practically all ions' energy is spent on the heating of the polymer surface. The elevated surface temperature causes rapid loss of surface wettability [32].

Finally, it is worth stressing again that high hydrophilicity, as indicated by a low water droplet contact angle, does not necessarily result in optimal adhesion. The formation of agglomerates of well-oxidized molecular fragments on the polymer surface explains the paradox. The fragments add to the wettability but are easily removed from the polymer surface, so the adhesion of any coating on a polymer, rich with molecular fragments, will be inadequate.

Author Contributions: Conceptualization, G.P. and M.M.; methodology, G.P. and M.M.; formal analysis, G.P.; investigation, M.M.; data curation, G.P.; writing—original draft preparation, G.P. and M.M.; writing—review and editing, G.P. and M.M.; visualization, G.P.; project administration, G.P.; funding acquisition, G.P. and M.M. All authors have read and agreed to the published version of the manuscript.

Funding: This research was funded by the Slovenian Research and Innovation Agency, national project L2-50078 "Nanocellulose from eco-farms for optimal enforcement of bioplastics" and national core funding P2-0082 "Thin-film structures and plasma surface engineering".

Data Availability Statement: This is a review article; therefore, no new data were generated.

Conflicts of Interest: The authors declare no conflicts of interest.

References

1. Awaja, F.; Gilbert, M.; Kelly, G.; Fox, B.; Pigram, P.J. Adhesion of polymers. *Prog. Polym. Sci.* **2009**, *34*, 948–968. [CrossRef]
2. Żołek-Tryznowska, Z.; Izdebska, J.; Gołąbek, M. Ionic liquids as performance additives for water-based printing inks. *Color. Technol.* **2014**, *130*, 314–318. [CrossRef]
3. von Keudell, A.; Schulz-von der Gathen, V. Foundations of low-temperature plasma physics—An introduction. *Plasma Sources Sci. Technol.* **2017**, *26*, 113001. [CrossRef]
4. Benedikt, J.; Kersten, H.; Piel, A. Foundations of measurement of electrons, ions and species fluxes toward surfaces in low-temperature plasmas. *Plasma Sources Sci. Technol.* **2021**, *30*, 033001. [CrossRef]
5. Ono, S.; Teii, S. Vibrational temperature in a weakly ionised steady-state nitrogen discharge plasma. *J. Phys. D Appl. Phys.* **1983**, *16*, 163–170. [CrossRef]
6. Liu, F.; Nie, L.; Lu, X.; Stephens, J.; Ostrikov, K. Atmospheric plasma VUV photon emission. *Plasma Sources Sci. Technol.* **2020**, *29*, 065001. [CrossRef]
7. Booth, J.-P.; Mozetič, M.; Nikiforov, A.; Oehr, C. Foundations of plasma surface functionalization of polymers for industrial and biological applications. *Plasma Sources Sci. Technol.* **2022**, *31*, 103001. [CrossRef]
8. Chytrosz-Wrobel, P.; Golda-Cepa, M.; Stodolak-Zych, E.; Rysz, J.; Kotarba, A. Effect of oxygen plasma-treatment on surface functional groups, wettability, and nanotopography features of medically relevant polymers with various crystallinities. *Appl. Surf. Sci. Adv.* **2023**, *18*, 100497. [CrossRef]
9. Fukunaga, Y.; Longo, R.C.; Ventzek, P.L.G.; Lane, B.; Ranjan, A.; Hwang, G.S.; Hartmann, G.; Tsutsumi, T.; Ishikawa, K.; Kondo, H.; et al. Interaction of oxygen with polystyrene and polyethylene polymer films: A mechanistic study. *J. Appl. Phys.* **2020**, *127*, 023303. [CrossRef]

10. Polito, J.; Denning, M.; Stewart, R.; Frost, D.; Kushner, M.J. Atmospheric pressure plasma functionalization of polystyrene. *J. Vac. Sci. Technol. A* **2022**, *40*, 043001. [CrossRef]
11. Bruggeman, P.J.; Iza, F.; Brandenburg, R. Foundations of atmospheric pressure non-equilibrium plasmas. *Plasma Sources Sci. Technol.* **2017**, *26*, 123002. [CrossRef]
12. Mirahmadi, M.; Pérez-Ríos, J. Three-body recombination in physical chemistry. *Int. Rev. Phys. Chem.* **2023**, *41*, 233–267. [CrossRef]
13. Li, Y.; Zhou, Y.; Gu, Y.; Chen, B.; Wang, B.; Yan, J.; Liu, J.; Chen, F.; Zhao, D.; Liu, X. Improving surface wettability and adhesion property of polytetrafluoroethylene by atmospheric-pressure ammonia water-mixed plasma treatment. *Vacuum* **2022**, *196*, 110763. [CrossRef]
14. Lojen, D.; Zaplotnik, R.; Primc, G.; Mozetič, M.; Vesel, A. Effect of VUV radiation and reactive hydrogen atoms on depletion of fluorine from polytetrafluoroethylene surface. *Appl. Surf. Sci.* **2020**, *533*, 147356. [CrossRef]
15. Mozetic, M.; Vesel, A.; Primc, G.; Zaplotnik, R. Non-thermal plasma technology for polymeric materials: Applications in composites nanostructured materials, and biomedical fields. In *Introduction to Plasma and Plasma Diagnostics*; Thomas, S., Mozetic, M., Cvelbar, U., Eds.; Elsevier: Amsterdam, The Netherlands, 2019; pp. 23–65.
16. Golda, J.; Biskup, B.; Layes, V.; Winzer, T.; Benedikt, J. Vacuum ultraviolet spectroscopy of cold atmospheric pressure plasma jets. *Plasma Process. Polym.* **2020**, *17*, 1900216. [CrossRef]
17. Fouchier, M.; Pargon, E.; Azarnouche, L.; Menguelti, K.; Joubert, O.; Cardolaccia, T.; Bae, Y.C. Vacuum ultra violet absorption spectroscopy of 193 nm photoresists. *Appl. Phys. A.* **2011**, *105*, 399–405. [CrossRef]
18. Antipova, V.; Omelyanchik, A.; Sobolev, K.; Pshenichnikov, S.; Vorontsov, S.; Korepanova, E.; Schitz, D.; Peddis, D.; Panina, L.; Levada, K.; et al. Enhancing wettability and adhesive properties of PVDF-based substrates through non-thermal helium plasma surface modification. *Polymer* **2024**, *290*, 126567. [CrossRef]
19. Takenaka, K.; Jinda, A.; Nakamoto, S.; Koyari, R.; Toko, S.; Uchida, G.; Setsuhara, Y. Influence of pre-treatment using non-thermal atmospheric pressure plasma jet on aluminum alloy A1050 to PEEK direct joining with hot-pressing process. *Int. J. Adv. Manuf. Technol.* **2023**, *130*, 1925–1933. [CrossRef]
20. Kosmachev, P.V.; Panin, S.V.; Panov, I.L.; Bochkareva, S.A. Structure and Deformation Behavior of Polyphenylene Sulfide-Based Laminates Reinforced with Carbon Fiber Tapes Activated by Cold Atmospheric Plasma. *Polymers* **2023**, *16*, 121. [CrossRef] [PubMed]
21. Kosmachev, P.V.; Panin, S.V.; Panov, I.L.; Bochkareva, S.A. Surface Modification of Carbon Fibers by Low-Temperature Plasma with Runaway Electrons for Manufacturing PEEK-Based Laminates. *Materials* **2022**, *15*, 7625. [CrossRef]
22. Jung, U.; Sang Kim, Y.; Suhr, J.; Lee, H.-s.; Kim, J. Enhancing adhesion strength via synergic effect of atmospheric pressure plasma and silane coupling agent. *Appl. Surf. Sci.* **2023**, *640*, 158227. [CrossRef]
23. Kim, Y.S.; Jeon, E.S. Effects of surface treatment method and fabric type on adhesion between PP material and artificial fabrics. *Surf. Interfaces* **2023**, *42*, 103464. [CrossRef]
24. Ondiek, W.; Kondo, M.; Adachi, M.; Macadre, A.; Goda, K. Effect of Surface Coating and Plasma Treatment on Mechanical Properties of Wood Plastic Composites. *J. Compos. Sci.* **2023**, *7*, 296. [CrossRef]
25. *JIS K 7139:2009*; Plastics—Test Specimens. Japanese Standards Association: Tokyo, Japan, 2009.
26. Gosar, Ž.; Đonlagić, D.; Pevec, S.; Gergič, B.; Mozetič, M.; Primc, G.; Vesel, A.; Zaplotnik, R. Distribution of the Deposition Rates in an Industrial-Size PECVD Reactor Using HMDSO Precursor. *Coatings* **2021**, *11*, 1218. [CrossRef]
27. Piskarev, M.S.; Shapran, Y.V.; Zinoviev, A.V.; Kechek'yan, A.S.; Gatin, A.K.; Gilman, A.B.; Kuznetsov, A.A. Contact and Adhesion Properties of Polyethylene Naphthalate Films Modified by DC Discharge. *Inorg. Mater. Appl. Res.* **2022**, *13*, 1398–1402. [CrossRef]
28. *ASTM 1876-2001*; Standard Test Method for Peel Resistance of Adhesives (T-Peel Test). ASTM International: West Conshohocken, PA, USA, 2001.
29. Gudmundsson, J.T.; Hecimovic, A. Foundations of DC plasma sources. *Plasma Sources Sci. Technol.* **2017**, *26*, 123001. [CrossRef]
30. Liu, X.; Zhen, G.; Chen, P. Effect of plasma treatment on high temperature interfacial properties of Chinese-made T800 CFRP. *Compos. Interfaces* **2023**, 1–14. [CrossRef]
31. Zaplotnik, R.; Primc, G.; Paul, D.; Mozetič, M.; Kovač, J.; Vesel, A. Atomic Species Generation by Plasmas. *Plasma Appl. Mater. Modif.* **2021**, 107–176. [CrossRef]
32. Oh, J.-H.; Moon, M.-W.; Park, C.H. Effect of crystallinity on the recovery rate of superhydrophobicity in plasma-nanostructured polymers. *RSC Adv.* **2020**, *10*, 10939–10948. [CrossRef]
33. Wang, R.; Liu, P.; Liu, J.; Wang, T.; Pang, X.; Xu, L.; Qiao, L. Enhancing the bond strength of magnetron-sputtered copper layers on polyimide and polyether ether ketone surfaces through surface modification. *J. Adhes. Sci. Technol.* **2024**, 1–17. [CrossRef]
34. Wu, Y.; Wang, T.; Gao, J.; Zhang, L.; Fay, J.D.B.; Hirth, S.; Hankett, J.; Chen, Z. Molecular Behavior of 1K Polyurethane Adhesive at Buried Interfaces: Plasma Treatment, Annealing, and Adhesion. *Langmuir* **2023**, *39*, 3273–3285. [CrossRef] [PubMed]
35. *ASTM D3136*; Standard Terminology Relating to Care Labeling for Apparel, Textile, Home Furnishing, and Leather Products. ASTM International: West Conshohocken, PA, USA, 2020.
36. Chabert, P.; Tsankov, T.V.; Czarnetzki, U. Foundations of capacitive and inductive radio-frequency discharges. *Plasma Sources Sci. Technol.* **2021**, *30*, 024001. [CrossRef]
37. Lin, T.-C.; Bruce, R.L.; Oehrlein, G.S.; Phaneuf, R.J.; Kan, H.-C. Direct and quantitative evidence for buckling instability as a mechanism for roughening of polymer during plasma etching. *Appl. Phys. Lett.* **2012**, *100*, 233113. [CrossRef]
38. Mozetic, M. Plasma-Stimulated Super-Hydrophilic Surface Finish of Polymers. *Polymers* **2020**, *12*, 2498. [CrossRef] [PubMed]

39. Wu, Y.; Wang, T.; Fay, J.D.B.; Zhang, L.; Hirth, S.; Hankett, J.; Chen, Z. Silane Effects on Adhesion Enhancement of 2K Polyurethane Adhesives. *Langmuir* **2023**, *39*, 19016–19026. [CrossRef] [PubMed]
40. Bhoj, A.N.; Kushner, M.J. Continuous processing of polymers in repetitively pulsed atmospheric pressure discharges with moving surfaces and gas flow. *J. Phys. D Appl. Phys.* **2007**, *40*, 6953–6968. [CrossRef]
41. Dorai, R.; Kushner, M.J. A model for plasma modification of polypropylene using atmospheric pressure discharges. *J. Phys. D Appl. Phys.* **2003**, *36*, 666–685. [CrossRef]
42. Longo, R.C.; Ranjan, A.; Ventzek, P.L.G. Density Functional Theory Study of Oxygen Adsorption on Polymer Surfaces for Atomic-Layer Etching: Implications for Semiconductor Device Fabrication. *Acs Appl. Nano Mater.* **2020**, *3*, 5189–5202. [CrossRef]
43. Kondeti, V.S.S.K.; Zheng, Y.; Luan, P.; Oehrlein, G.S.; Bruggeman, P.J. O·, H·, and ·OH radical etching probability of polystyrene obtained for a radio frequency driven atmospheric pressure plasma jet. *J. Vac. Sci. Technol. A* **2020**, *38*, 033012. [CrossRef]
44. Strobel, M.; Jones, V.; Lyons, C.S.; Ulsh, M.; Kushner, M.J.; Dorai, R.; Branch, M.C. A Comparison of Corona-Treated and Flame-Treated Polypropylene Films. *Plasmas Polym.* **2003**, *8*, 61–95. [CrossRef]
45. Arikan, E.; Holtmannspötter, J.; Zimmer, F.; Hofmann, T.; Gudladt, H.-J. The role of chemical surface modification for structural adhesive bonding on polymers—Washability of chemical functionalization without reducing adhesion. *Int. J. Adhes. Adhes.* **2019**, *95*, 102409. [CrossRef]
46. Vida, J.; Ilčíková, M.; Přibyl, R.; Homola, T. Rapid Atmospheric Pressure Ambient Air Plasma Functionalization of Poly(styrene) and Poly(ethersulfone) Foils. *Plasma Chem. Plasma Process.* **2021**, *41*, 841–854. [CrossRef]
47. Jokinen, V.; Suvanto, P.; Franssila, S. Oxygen and nitrogen plasma hydrophilization and hydrophobic recovery of polymers. *Biomicrofluidics* **2012**, *6*, 016501. [CrossRef]
48. Vrlinic, T.; Mille, C.; Debarnot, D.; Poncin-Epaillard, F. Oxygen atom density in capacitively coupled RF oxygen plasma. *Vacuum* **2009**, *83*, 792–796. [CrossRef]
49. Vesel, A.; Zaplotnik, R.; Primc, G.; Mozetič, M. Evolution of surface functional groups and aromatic ring degradation upon treatment of polystyrene with hydroxyl radicals. *Polym. Degrad. Stab.* **2023**, *218*, 110582. [CrossRef]
50. Vesel, A.; Zaplotnik, R.; Primc, G.; Mozetič, M. Evolution of the surface wettability of PET polymer upon treatment with an atmospheric-pressure plasma jet. *Polymers* **2020**, *12*, 87. [CrossRef] [PubMed]
51. Vesel, A.; Zaplotnik, R.; Mozetič, M.; Primc, G. Surface modification of PS polymer by oxygen-atom treatment from remote plasma: Initial kinetics of functional groups formation. *Appl. Surf. Sci.* **2021**, *561*, 150058. [CrossRef]
52. Wertheimer, M.R.; Fozza, A.C.; Holländer, A. Industrial processing of polymers by low-pressure plasmas: The role of VUV radiation. *Nucl. Instrum. Methods Phys. Res. Sect. B Beam Interact. Mater.* **1999**, *151*, 65–75. [CrossRef]
53. Fantz, U.; Briefi, S.; Rauner, D.; Wünderlich, D. Quantification of the VUV radiation in low pressure hydrogen and nitrogen plasmas. *Plasma Sources Sci. Technol.* **2016**, *25*, 045006. [CrossRef]

materials

MDPI

Article

Study on the Printability of Starch-Based Films Using Ink-Jet Printing

Zuzanna Żołek-Tryznowska [1,*], Katarzyna Piłczyńska [1], Tomasz Murawski [1], Arkadiusz Jeznach [1] and Krzysztof Niczyporuk [2]

[1] Faculty of Mechanical and Industrial Engineering, Warsaw University of Technology, Narbutta 85, 02-524 Warsaw, Poland; katarzyna.pilczynska@pw.edu.pl (K.P.); tomasz.murawski.dokt@pw.edu.pl (T.M.); arkadiusz.jeznach@pw.edu.pl (A.J.)

[2] Association of Polish Engineers and Mechanical Technicians, Czackiego 3/5, 00-043 Warsaw, Poland; krzysztof.niczyporuk@simp.pl

* Correspondence: zuzanna.tryznowska@pw.edu.pl

Abstract: Starch-based films are a valuable alternative to plastic materials that are based on fossil and petrochemical raw resources. In this study, corn and potato starch films with 50% glycerol as a plasticizer were developed, and the properties of films were confirmed by mechanical properties, surface free energy, surface roughness, and, finally, color and gloss analyses. Next, the films were overprinted using ink-jet printing with quick response (QR) codes, text, and pictograms. Finally, the print quality of the obtained prints was determined by optical density, color parameters, and the visual evaluation of prints. In general, corn films exhibit lower values of mechanical parameters (tensile strength, elongation at break, and Young Modulus) and water transition rate (11.1 mg·cm^{-2}·h^{-1}) than potato starch film (12.2 mg·cm^{-2}·h^{-1}), and water solubility is 18.7 ± 1.4 and 20.3 ± 1.2% for corn and potato film, respectively. The results obtained for print quality on starch-based films were very promising. The overprinted QR codes were quickly readable by a smartphone. The sharpness and the quality of the lettering are worse on potato film. At the same time, higher optical densities were measured on potato starch films. The results of this study show the strong potential of using starch films as a modern printing substrate.

Keywords: starch films; ink-jet printing; print quality; QR code; packaging

1. Introduction

There is a growing demand for ecologically sustainable alternatives to traditional plastic packaging. The researchers seek novel, functional, environmentally friendly, or biobased materials. The research on biodegradable and renewable materials follows the UE directive [1]. Starch is abundant, biodegradable, biocompatible, cheap, and renewable. Also, it should be highlighted that it can be harvested without destroying the plant [2,3]. Due to their excellent filmogenic properties, low cost, and biodegradation, thermoplastic starch-based films are great alternatives for plastics that are based on fossil and petrochemical raw materials. The number of research papers in the field of starch-based materials for packaging purposes is gradually increasing. The starch-based composite or blend has great potential for commercialization in the food packaging field due to it being pollution-free and its simple preparation, desirable functionality, degradability, and low digestibility [4].

Unfortunately, starch materials exhibit some disadvantages that limit their usage as packaging films: moisture sensitivity and low mechanical strength and stability [5]. To improve the properties of thermoplastic starch films, the starch can be modified prior to film preparation by combining it with other biodegradable materials or reinforcing it with active or natural nanoparticles. The enzymatic hydrolysis using α-amylase from *Bacillus amyloliquefaciens* can be used successfully to improve corn starch-based film's mechanical properties, such as greater elasticity, strength, and stiffness [6]. Adding various natural extracts improves the mechanical properties of the starch-based films, i.e., *Thymus vulgaris* essential oils (TEOs) combined with ethanolic extract of propolis (EEP) increases the elongation at break up to 13.5% [7]. The combination of starch with other polymers, such as polyvinyl alcohol, improves the mechanical and barrier properties of film packaging [8–10]. Cellulose nanofibers and nanocrystals can be successfully applied to enhance starch-based materials' properties [11]. Zeleke et al. have reported the usage of rice straw cellulose fibers to reinforce thermoplastic starch-based films for better industrial applications [12]. According to [12], 10 wt% cellulose nanofibers were the optimum concentration for the composite film, compromising the properties and the transparency. Above 10 wt%, the fiber aggregation in the polymer matrix was observed. Finally, blending plasticized starch with low-density polyethylene allows material development with greater mechanical properties and improved biodegradation for effective environmental pollution control [13].

The packaging materials must fulfill some criteria. The primary function of packaging is to protect the product against external factors. However, the packaging should also communicate accurate information through printed texts and graphics on relevant features of the product, such as ingredients, weight, etc. [14]. Good transparency allows the consumer to see the product packed inside the packaging. Moreover, transparency can be a desired factor for film printing and a legible inscription reading. Therefore, apart from the basic properties of starch films, such as mechanical, barrier, surface, etc., the possibility of printing is crucial. Our previous work showed the potential of overprinting starch films with water-based inks for flexographic technology [15]. The packaging can be overprinted with various techniques, including flexographic, digital, and many more.

Today, ink-jet printing technology is one of the simplest and most popular digital printing techniques, allowing printing on the scale of "short runs" [16]. In the ink-jet printing process, there is a controlled ejection of the liquid on the substrate in a very precise and controlled way [17]. The printing process is contactless, so there is no need for an intermediate carrier to transfer the ink to the substrate. There is no limitation on the substrate type—it can be flexible or rigid—so this technique is suitable for printing on many substrates, including paper, cardboard, textiles, films, glass, metals, etc. In addition, ink-jet printing enables the usage of many different colors and large-format prints. Ink-jet printing has been found to be a widespread technology with broad applications: printing on textiles, production of printing plates, and, finally, functional printing such as 3D printing or printing of so-called printed electronics [18–22]. Ink-jet does not require the usage of expensive printing forms or screens. After a print, the design can be quickly sent to a printing machine and printed on various substrates.

In theory, ink-jet is simple—a print head ejects tiny drops of ink onto the substrate [19]. In practice, ink-jet printing is a complex problem, where implementation of the ink-jet technology needs multidisciplinary skills. The final print quality is influenced by several parameters, mainly the properties of the ink, the substrate, and the printing process parameters. The properties of the ink, such as viscosity and rheology, influence the ink flow in the printhead [16,23]. Next, the surface tension, ink flow, and viscosity influence the drop formation in the printhead [24]. Finally, the properties of the printing substrate affect the final print quality [25,26].

A quick response (QR) code can be used as a data carrier for smart packaging. A QR code is a 2D variant of a barcode, which encodes valuable information. The QR code can be generated simply by using various free tools. A sample QR code (see Figure 1a) was generated using Adobe Express's QR code creator [27]. The typical QR code is composed of black and white blocks, where they refer to 1 and 0, respectively, representing the bit of information [28]. The white and black squares form a square-shaped grid. Three large squares in the corner allow position detection, size, angle, and shape recognition [29]. This kind of QR code has a poor visual effect. However, by using an appropriate scanning algorithm, a more aesthetic, colorful, and visually pleasant QR code can be generated [30], as shown in Figure 1b–d.

(a) (b) (c) (d)

Figure 1. Samples of QR codes, (**a**) basic QR code with conventional square blocks; (**b–d**) QR codes with aesthetic blocks.

The QR code can be scanned using a scanner or a smartphone with an installed QR code scanner app. The QR code can code useful information about the product in the packaging. QR codes are widely used on a daily basis, and they have been successfully adopted in mobile payments [28], in logistics sorting to allow package tracking and sorting [30], in in situ air analyses [29], and, finally, in food traceability systems to obtain information on the freshness and quality of food products [31–33]. Moreover, a QR code can be used to deliver a known amount of a drug [34].

Smart packaging can be developed by combining various printing techniques, materials, and QR codes. An individual QR code label is in a large-scale, rapid, low-cost fabrication process, whereas a unique QR code is overprinted, i.e., one item–one code [35].

This study aims to analyze the print quality of overprinted starch-based films. We have developed two kinds of starch films based on corn and potato starch as a printing base for developing smart, biodegradable packaging to achieve this goal. The properties of starch films were confirmed through mechanical, surface, water vapor transition rate, water solubility, and film color analyses. The developed materials were overprinted with an ink-jet technique with various pictograms and QR codes. Finally, the print quality of the prints was assessed. The starch material with an overprinted QR code giving information about the product forms a unique, smart, and biodegradable packaging for food.

2. Materials and Methods

2.1. Materials

Corn (NaturAvena, Piaseczno, Poland) and potato (Bio Planet, Leszno, Poland) food-grade starches were used. Glycerol (purity $\geq$ 99%, CAS 56-81-5) and diiodomethane (purity $\geq$ 99%, CAS 75-11-6) were purchased from Sigma-Aldrich (Poznań, Poland) and used as received. For the contact angle measurement, water was purified by electrodeionization with the MilliporeSigma Elix Water Purification System (Burlington, MA, USA).

2.2. Film Preparation

We previously showed the procedure in our works, for example, see [36]. The reagents (10 g of starch, 5 g of glycerol in 200 g of water) were heated up to 95 °C upon stirring with a mechanical stirrer and cast onto Teflon® plates placed on a K Paint Applicator (RK Print, Royston, UK) equipped with an adjustable micrometer spreader gap set to 3 mm with a constant coating speed (6 m·min^{-1}). The starch films were dried in the climate room for one week in controlled conditions (23 $\pm$ 0.5 °C; 50 $\pm$ 1% RH).

2.3. Film Properties Determination

ATR FT-IR spectra were recorded at room temperature in the 400–4000 cm^{-1} range with a resolution of 4 cm^{-1}, using a Nicolet iS5 spectrometer equipped with a Platinum single-reflection diamond ATR module. The FT-IR spectra were analyzed with OMNIC Spectra™ software (series 9.12.968).

The mechanical properties were determined by using a Z010 tensiometer (Zwick-Roell, Ulm, Germany). The measurement was performed according to the ISO 527-1 standard [37]: 15 mm in width, 100 mm in length; the initial distance of the clamps was 50 mm; the stretching speed was 100 mm·m^{-1}. The thickness for the mechanical measurements was performed with a handheld micrometer, with a 0.001 mm resolution and error of ±0.5 mm. The parameters (tensile strength, elongation at break, and Young's Modulus) were obtained from the stress–strain curves. The measurements were repeated ten times, and the average values were taken as the result.

The surface free energy was calculated using the Owens–Wendt–Rabel–Kaelbe approach [38], based on the contact angle measurement results of water and diiodomethane. The static contact angle was measured with the Drop Shape Analysis System (DSA 30E, Krüss, Hamburg, Germany) in agreement with the ISO 15989 standard [39].

A digital microscope (Keyence VHX-7000, Keyence Corporation, Osaka, Japan) was used to observe the surface structure and the surface roughness determination of developed films. The film's linear roughness (R_a and R_z) was subsequently evaluated on the Keyence VHX-7000.

The water vapor transmission rate (WVTR) measurements were performed using the MA 210.R (Radwag, Radom, Poland) moisture analyzer. A 54 mm ± 2 mm diameter film sample was placed on an aluminum-sealed probe with 5 g of water and weighed. The WTVR was determined at 45 °C at constant room temperature (23 ± 0.5 °C) and relative humidity (50 ± 1% RH). The measurements were performed manually, and the samples were weighed at 0, 1, and 2 h intervals. The WVTR was calculated using Equation (1):

$$WVTR = \frac{\Delta m}{t \cdot S} \tag{1}$$

where Δm is the difference between the mass of water after 2 and 1 h of measurement, t is the time of measurement (2 h), and S is the area of the film sample.

The Karl Fisher volumetric titration method (Mettler V30, Beersel, Belgium) was applied for the determination of water content. The sample (approx. 200 mg) was added to a sealed vial. Two component reagents Aquastar®-Solvent and Aquastar®-Titrant 2 (Supelco, Sigma-Aldrich, Poznań, Poland) were used as a solvent and titrant, respectively.

The solubility of starch-based films was determined according to [40]. Samples of films (approx. 20 mm × 20 mm) were solubilized for 24 h in 50 mL of distilled water (at 25 °C). The solubility was expressed as the percent weight loss of the film strips on soaking, according to Equation (2):

$$Solubility\ (\%) = \left(\frac{m_i - m_f}{m_i}\right) \cdot 100 \tag{2}$$

where m_i is the initial dry mass of the sample and m_f is the final dry mass of the sample after solubilization.

The measurement of WVTR, solubility, and moisture content was repeated in triplicate.

The gloss (in the gloss units, GU) of the prints was measured at 20°, 60°, and 85° geometry conditions with the use of a Micro-Tri-Gloss gloss meter (BYK-Gardner, Geretsried, Germany). Data collection was performed at six different positions of the samples in both directions: cross and machine direction. The reported values are the average of these measurements.

The color of the films was measured using an X-Rite eXact spectrophotometer (Grand Rapids, MI, USA) under the following conditions: D50 luminant, 2° colorimetric observer, and M2 (UVC) filter. The white ink-jet paper was used as the white standard (L^* = 93.73, a^* = 0.79, and b^* = −3.98). The measurement was repeated three times.

2.4. Printing

The printing was performed with an eZcolorJet (Graph-Tech, Ft. Pierce, FL, USA) industrial printer equipped with an EPSON I3200 printhead. The printing was performed at 600 dpi with magenta and black UV LED ink (Graph-Tech, Ft. Pierce, FL, USA) printing with 6 pl per 1 drop with max. 42 kHz jetting.

2.5. Print Quality Analyses

Optical densities and color values were determined with a SpectroEye spectrophotometer (GretagMacbeth, Zürich, Switzerland). The measurement was performed with the following settings: D50 luminant, 2° colorimetric observer, without polarization filter. The measurement was repeated three times.

The prints were evaluated with a digital portable microscope model Dino-Lite AM 4113T-FVW Premier Handheld Digital Microscope (Dino-Lite Europe, Almare, The Netherlands), to assess print quality, edge sharpness, and microlevel non-uniformity. The selected parts of the prints were captured with 200× magnification and 1280 × 1024 resolution, using DinoCapture 2.0 v 1.5.48 (Dino-Lite Europe, Almare, The Netherlands).

A QR scanning application on a smartphone was used to scan the QR code. In this study, a Mobile Barcode Scanner (v. 5.0.9 (175), Cognex, Natick, MA, USA) and a camera app (Galaxy S21, Samsung, Seoul, Republic of Korea) were used.

All the results are presented as a mean ± SD, where a mean was calculated as an average and SD as a standard deviation using Excel (ver. 2312, Office 365, Microsoft, Redmond, WA, USA).

3. Results and Discussion

3.1. Properties of Films

Starch films based on potato and corn with 50% glycerol as plasticizers were developed as a printing base. To confirm the quality of starch films, the basic properties were determined, i.e., mechanical properties, surface free energy, water solubility, etc. The measured properties are summarized in Table 1. The mechanical properties, water vapor transition rate, and water solubility determine the use of starch-based materials for packaging purposes. Surface-free energy is the crucial factor influencing the printability of starch-based films. Figure 2 shows the FTIR spectra of both films and the image of the films' surface.

The ATR-FTIR analyses show no changes in absorption bands, whether corn or potato starch was used for the film development (see Figure 1a). The spectrum shows typical peak characteristics for starch materials: broadband at approx. 3300 cm^{-1}, corresponding to O–H stretching vibrations in the hydroxyl group at approx. 2900 cm^{-1}, a characteristic peak for CH_2 stretching vibration, and at approx. 1020 cm^{-1}, characteristic of C–O–C stretching vibration [41].

Table 1. Properties of developed films.

Property		Starch Film	
		Corn	Potato
Young's Modulus (MPa)		42.6 ± 1.2	53.4 ± 1.6
Tensile strength (MPa)		3.47 ± 0.68	4.05 ± 0.60
Elongation at break (%)		33 ± 13	43 ± 6
Thickness (μm)		201 ± 7	80 ± 8
SFE ($mJ \cdot m^{-2}$)		58.67 ± 4.28	59.71 ± 2.39
Polar component of SFE ($mJ \cdot m^{-2}$)		18.89 ± 2.38	17.02 ± 1.15
Surface roughness, R_a (μm)		1.92 ± 0.16	1.27 ± 0.69
WVTR (%); ($mg \cdot cm^{-2} \cdot h^{-1}$)		39; 11.1	43; 12.2
Water content (%)		7.16 ± 0.57	6.48 ± 0.41
Water solubility (%)		18.7 ± 1.4	20.3 ± 1.2
Gloss (°)		4.1 ± 0.9	11.0 ± 3.1
Color	L^*	90.64 ± 0.23	90.72 ± 0.40
	a^*	0.52 ± 0.01	0.84 ± 0.01
	b^*	−2.24 ± 0.08	−3.60 ± 0.04

Values are means ± SD.

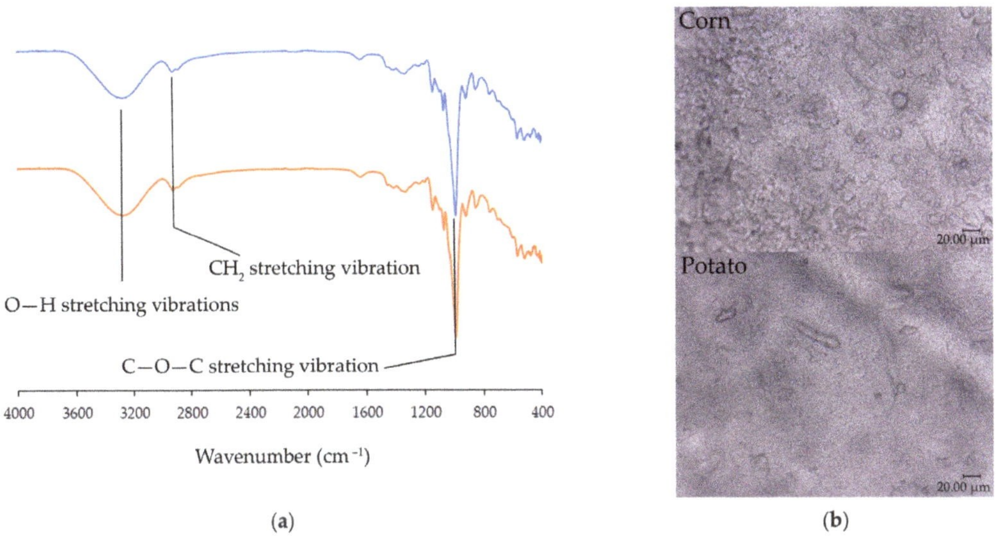

Figure 2. (**a**) FTIR spectrum of corn (blue line) and potato (orange line) film; (**b**) surface images of films at 900× magnification.

The tensile strength and elongation values at break are comparable to those reported in our previous work [42] and higher than those reported by Dai et al. [43]. The values of mechanical parameters were lower for corn than potato starch films.

The surface roughness, together with the value of surface free energy (SFE) and the polar component of SFE, indicates the possibility of printing. The values of SFE and the polar component of SFE for corn and potato starch films are comparable. At the same time, surface roughness, R_a, is slightly higher for corn starch film.

The barrier properties of films indicate their packaging application. In general, the hydrophilic nature of starch films plasticized with glycerol influences the water vapor permeability [44]. High affinity of glycerol to water promotes the diffusion of water molecules through the films [45]. Potato starch exhibits higher values of WVTR and higher water solubility.

The values of gloss and color properties such as lightness (L^*) and hue parameters (a^*, b^*) are listed in Table 1. For the gloss measurement, geometry with an angle of 85° was used because the gloss of starch materials exhibits lower values than 10 gloss units using a geometry of 60° [46].

The L^* values are similar for corn and potato films. It should be noted that when comparing potato and corn films, both the a^* value and the b^* value for the potato became smaller. Hence, the potato starch is "greener" and "bluer" compared to corn starch film. The color difference between both starch films can be observed with the naked human eye, which is confirmed by the value of the color difference (calculated from the equation: $\Delta E^*_{ab} = \sqrt{(\Delta L^*)^2 + (\Delta a^*)^2 + (\Delta b^*)^2}$, where ΔL^*, Δa^*, and Δb^* are the differences between the values of color parameters on corn and potato film), $\Delta E = 1.4$.

3.2. Printing and Print Quality Assessment

The "print quality" has no strict definition [47], and the assessment depends on the customer's expectations. The quality of the prints can be assessed by spectrophotometric measurements of color coordinates, the optical density of full-tone coverage, gloss measurement, and adhesion tests [48]. This study determined print quality by determining color parameters L^*, a^*, b^*, and optical densities. The naked human eye control reveals that the prints on the potato starch film are of poorer quality than the overprinted corn or paper samples. The overprinted QR codes and pictograms are shown in Figure 3. The prints were also visually evaluated (Figures 4 and 5).

(a) (b) (c)

Figure 3. (**a**) Overprinted pictograms on starch-based film; (**b**) overprinted QR-codes; (**c**) screenshot of a smartphone after scanning the QR code.

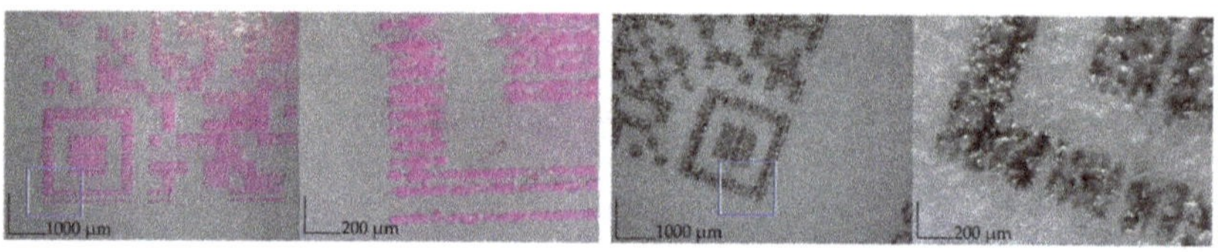

(a) (b)

Figure 4. The printed QR codes with magenta and black ink on corn starch: overprinted with magenta ink at magnification 50× and 225× (**a**); overprinted with black ink at magnification 50× and 225× (**b**).

The print on the paper substrate was used as a reference. Images of the standard QR code and selected pictograms were printed with an ink-jet printer with black and magenta ink. The QR code was printed directly on the surface of the starch film and was scanned using a QR code scanner app installed on a smartphone. The QR code image's scale, color, and quality critically influence the possibility of scanning. To analyze the deformation of the prints, the digital microscope was used to capture the selected areas on the printouts. To study the difference between images, specific areas on images were selected. Figure 4 shows the deformations of a QR code overprinted on corn starch with magenta and black ink. The overprinted QR code (Figure 3b) was read easily by a smartphone, as shown in Figure 3c. The brighter and less clear QR code block elements (see Figure 4) did not negatively affect its readability. This result was expected because QR codes can still be scanned even with a 30% error [49].

Figure 5 compares the quality of reproduction of printed images and fonts. The letter 'R' and the arrow from the compostable logo were analyzed and printed; line width was measured for the image analyses. As it is shown in Figure 4b, the line width depends on the printing substrate, and it is narrower on corn film. The width of the overprinted lines on the starch films is close to the values measured on the paper substrate. The properties and composition of the substrate influence the printed line width [50], which is related to the penetration of the substrate by the ink [51]. Furthermore, it is evident that the ink spread on potato film; therefore, the sharpness and letter quality are worse on potato film.

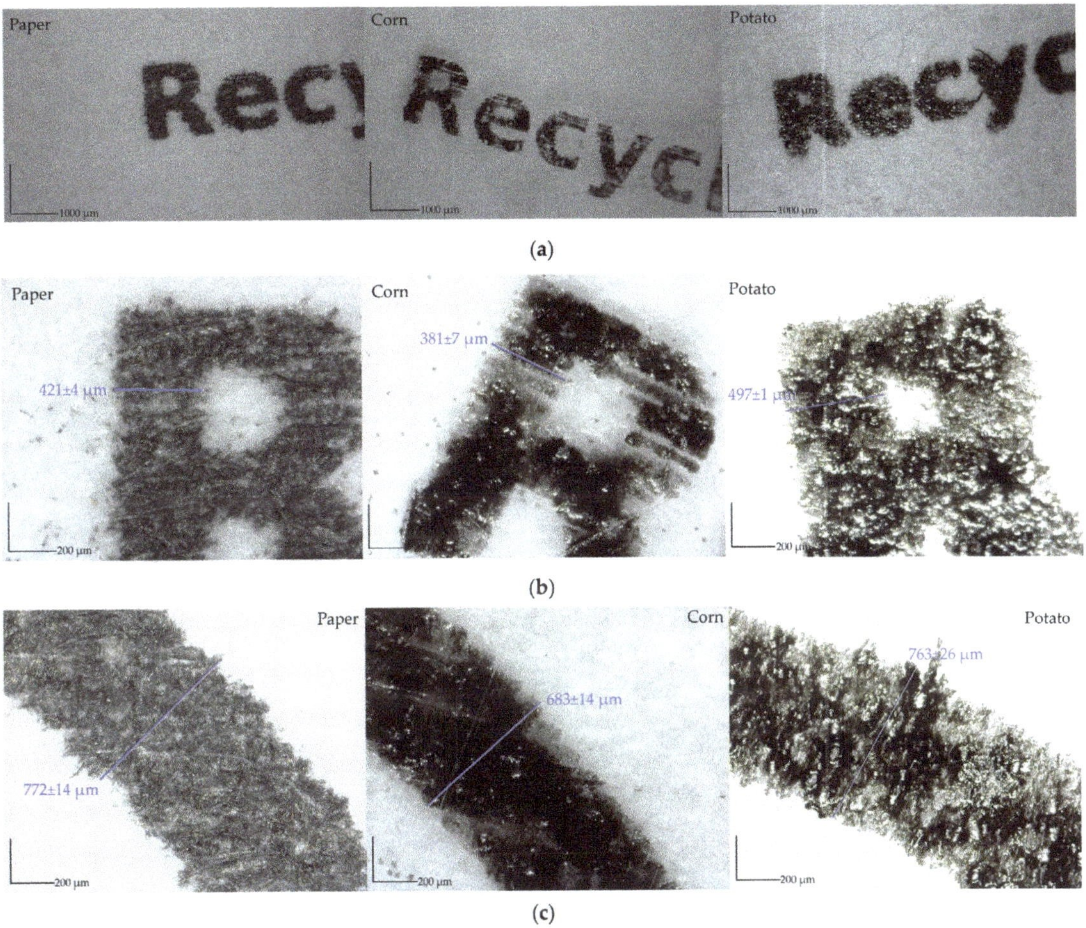

Figure 5. (**a**) Appearance of specific area at 50× magnification; (**b**) appearance of 'R' letter in word 'Recyclable' at 250× magnification; (**c**) appearance of arrow in word compostable logo at 250× magnification.

Table 2 summarizes the L^*, a^*, b^* color parameters and optical density (OD) values. The print color can be assessed by measuring L^*, a^*, b^* (lightness L^*, green–red coordinate a^*, and blue–yellow coordinate b^*). The value of L^* is lower for the overprinted potato and corn starch film than for the overprinted paper. This indicates that the print on starch films is darker than on paper. Moreover, the values of the a^* and b^* parameters are higher.

Table 2. Specific ink color components, L^*, a^*, b^*, and optical densities, *OD*, of ink layers printed on corn film.

Sample	L^*	a^*	b^*	*OD*
Paper	40.52 ± 1.31	0.80 ± 0.04	2.08 ± 0.26	0.92 ± 0.02
Corn	21.14 ± 2.78	2.05 ± 0.26	4.29 ± 0.80	1.46 ± 0.06
Potato	16.17 ± 1.81	2.23 ± 0.35	4.08 ± 0.72	1.61 ± 0.09

The optical density of overprinted starch-based films varied from 1.46 to 1.61. The similar values are comparable with the results determined for starch material [15] or other biodegradable polymers, such as PLA [52]. Higher optical density means higher print quality. It is visible that higher density was achieved when printing on the potato film; this may be related to the films' properties, such as surface free energy and surface roughness. Lower surface roughness and/or higher values of surface free energy of the substrate are related to a higher density due to homogenous ink film formation without the formation of a thicker amount of ink staying in the micro valleys of the substrate surface [53]. Furthermore, the color parameters (L^*, a^*, b^*) exhibited lower values than those for overprinted.

4. Conclusions

Printability is one of the significant functional properties of modern materials for packaging purposes. This paper was aimed at the printing performance of starch-based films as a packaging material. To evaluate the printing properties of starch films, corn, and potato starch films with 50% glycerol as a plasticizer were developed. The basic properties of films were measured, i.e., mechanical properties, surface free energy, water solubility, color gloss, etc.

The developed films were used as printing bases and overprinted with ink-jet in technological conditions. The basic print quality of films was assessed with image analyses, color parameters, and optical density. The print quality of overprinted QR codes allows for scanning with a smartphone. Furthermore, the overprinted text and pictograms are legible to the naked human eye. The sharpness and the quality of the lettering are worse on potato film. The line width of selected overprinted elements was wider on potato starch films. At the same time, higher optical densities were measured on potato starch films (1.61 $\pm$ 0.06) than on corn starch films (1.46 $\pm$ 0.06).

The findings of this study confirm that starch-based films can be used as modern, environmentally friendly packaging and can be overprinted with good quality. The results shown in this paper continue our previous works on the printing and packaging performance of starch-based materials. Our results expand the research on starch-based materials as a source of packaging material. The possibility of good-quality ink-jet printing supports the use of starch-based films for packaging, which makes visible necessary information regarding the packed product.

Author Contributions: Conceptualization, Z.Ż.-T. and K.N.; methodology, Z.Ż.-T. and K.N.; investigation, K.P., A.J., T.M. and K.N.; resources, Z.Ż.-T. and K.N.; writing—original draft preparation, Z.Ż.-T.; writing—review and editing, Z.Ż.-T.; visualization, Z.Ż.-T.; supervision, Z.Ż.-T. All authors have read and agreed to the published version of the manuscript.

Funding: This research received no external funding.

Institutional Review Board Statement: Not applicable.

Informed Consent Statement: Not applicable.

Data Availability Statement: The data presented in this study are available upon request from the corresponding author.

Conflicts of Interest: The authors declare no conflicts of interest.

References

1. The European Parliament and the Council of the European Union. Directive (EU) 2018/2001 on Promotion of the Use of Energy from Renewable Sources. 2018. Available online: https://eur-lex.europa.eu/eli/dir/2018/2001/oj (accessed on 17 January 2024).
2. Ray, R.; Narayan Das, S.; Das, A. Mechanical, Thermal, Moisture Absorption and Biodegradation Behaviour of Date Palm Leaf Reinforced PVA/Starch Hybrid Composites. *Mater. Today Proc.* **2021**, *41*, 376–381. [CrossRef]
3. Ojogbo, E.; Ogunsona, E.O.; Mekonnen, T.H. Chemical and Physical Modifications of Starch for Renewable Polymeric Materials. *Mater. Today Sustain.* **2020**, *7–8*, 100028. [CrossRef]
4. Su, C.Y.; Li, D.; Wang, L.; Wang, Y. Biodegradation Behavior and Digestive Properties of Starch-Based Film for Food Packaging—A Review. *Crit. Rev. Food Sci. Nutr.* **2023**, *63*, 6923–6945. [CrossRef] [PubMed]
5. Roy, S.; Rhim, J.W. Starch/Agar-Based Functional Films Integrated with Enoki Mushroom-Mediated Silver Nanoparticles for Active Packaging Applications. *Food Biosci.* **2022**, *49*, 101867. [CrossRef]

6. Ghizdareanu, A.-I.; Banu, A.; Pasarin, D.; Ionita, A.; Nicolae, C.-A.; Gabor, A.R.; Pătroi, D. Enhancing the Mechanical Properties of Corn Starch Films for Sustainable Food Packaging by Optimizing Enzymatic Hydrolysis. *Polymers* **2023**, *15*, 1899. [CrossRef] [PubMed]
7. Ardjoum, N.; Chibani, N.; Shankar, S.; Salmieri, S.; Djidjelli, H.; Lacroix, M. Incorporation of Thymus Vulgaris Essential Oil and Ethanolic Extract of Propolis Improved the Antibacterial, Barrier and Mechanical Properties of Corn Starch-Based Films. *Int. J. Biol. Macromol.* **2023**, *224*, 578–583. [CrossRef] [PubMed]
8. Gómez-Aldapa, C.A.; Velazquez, G.; Gutierrez, M.C.; Rangel-Vargas, E.; Castro-Rosas, J.; Aguirre-Loredo, R.Y. Effect of Polyvinyl Alcohol on the Physicochemical Properties of Biodegradable Starch Films. *Mater. Chem. Phys.* **2020**, *239*, 122027. [CrossRef]
9. Rahmani, H.; Mahmoudi Najafi, S.H.; Ashori, A.; Arab Fashapoyeh, M.; Aziz Mohseni, F.; Torkaman, S. Preparation of Chitosan-Based Composites with Urethane Cross Linkage and Evaluation of Their Properties for Using as Wound Healing Dressing. *Carbohydr. Polym.* **2020**, *230*, 115606. [CrossRef]
10. Fabijański, M.; Garbarski, J. Mechanical Properties of Thermoplastic Starch Filled with Calcium Carbonate. *Przem. Chem.* **2023**, *102*, 829–833. [CrossRef]
11. Lomelí-Ramírez, M.G.; Reyes-Alfaro, B.; Martínez-Salcedo, S.L.; González-Pérez, M.M.; Gallardo-Sánchez, M.A.; Landázuri-Gómez, G.; Vargas-Radillo, J.J.; Diaz-Vidal, T.; Torres-Rendón, J.G.; Macias-Balleza, E.R.; et al. Thermoplastic Starch Biocomposite Films Reinforced with Nanocellulose from Agave Tequilana Weber Var. Azul Bagasse. *Polymers* **2023**, *15*, 3793. [CrossRef]
12. Zeleke, T.S.; Yihun, F.A.; Ayana, M.T.; Kassa, M.T.; Alemante, M.F. Enhancing the Thermo-Mechanical Properties of Thermoplastic Starch Films Using Rice Straw Fibers as Reinforcement. *Chem. Afr.* **2023**, *6*, 2321–2329. [CrossRef]
13. Nwuzor, I.C.; Oyeoka, H.C.; Nwanonenyi, S.C.; Ihekweme, G.O. Biodegradation of Low-Density Polyethylene Film/Plasticized Cassava Starch Blends with Central Composite Design for Optimal Environmental Pollution Control. *J. Hazard. Mater. Adv.* **2023**, *9*, 100251. [CrossRef]
14. Luo, X.; Zaitoon, A.; Lim, L.T. A Review on Colorimetric Indicators for Monitoring Product Freshness in Intelligent Food Packaging: Indicator Dyes, Preparation Methods, and Applications. *Compr. Rev. Food Sci. Food Saf.* **2022**, *21*, 2489–2519. [CrossRef] [PubMed]
15. Żołek-Tryznowska, Z.; Holica, J. Starch Films as an Environmentally Friendly Packaging Material: Printing Performance. *J. Clean. Prod.* **2020**, *276*, 124265. [CrossRef]
16. Hussain, Z.; Kiaee, Z.; Nazarzadeh, M.; Reichel, C.; Tepner, S.; Tuladhar, T.; Jahn, M.; Keding, R. High-Frequency Rheological and Piezo-Voltage Waveform Characterization of Inkjet-Printed Polymer-Based Dopant-Source Inks. *Micromachines* **2022**, *14*, 80. [CrossRef] [PubMed]
17. Idbella, M.; Giusti, D.; Gulli, G.; Bonanomi, G. Structure, Functionality, Compatibility with Pesticides and Beneficial Microbes, and Potential Applications of a New Delivery System Based on Ink-Jet Technology. *Sensors* **2023**, *23*, 3053. [CrossRef]
18. Zub, K.; Hoeppener, S.; Schubert, U.S. Inkjet Printing and 3D Printing Strategies for Biosensing, Analytical, and Diagnostic Applications. *Adv. Mater.* **2022**, *34*, 2105015. [CrossRef]
19. Yang, P.; Zhang, L.; Kang, D.J.; Strahl, R.; Kraus, T. High-Resolution Inkjet Printing of Quantum Dot Light-Emitting Microdiode Arrays. *Adv. Opt. Mater.* **2020**, *8*, 1901429. [CrossRef]
20. Luo, X. Application of Inkjet-Printing Technology in Developing Indicators/Sensors for Intelligent Packaging Systems. *Curr. Opin. Food Sci.* **2022**, *46*, 100868. [CrossRef]
21. Arora, E.K.; Sharma, V.; Ravi, A.; Shahi, A.; Jagtap, S.; Adhikari, A.; Dash, J.K.; Kumar, P.; Patel, R. Polyaniline-Based Ink for Inkjet Printing for Supercapacitors, Sensors, and Electrochromic Devices. *Energies* **2023**, *16*, 6716. [CrossRef]
22. Lemarchand, J.; Bridonneau, N.; Battaglini, N.; Carn, F.; Mattana, G.; Piro, B.; Zrig, S.; Noël, V. Challenges, Prospects, and Emerging Applications of Inkjet-Printed Electronics: A Chemist's Point of View. *Angew. Chem. Int. Ed.* **2022**, *61*, e202200166. [CrossRef]
23. Dybowska-Sarapuk, L.; Kielbasinski, K.; Arazna, A.; Futera, K.; Skalski, A.; Janczak, D.; Sloma, M.; Jakubowska, M. Efficient Inkjet Printing of Graphene-Based Elements: Influence of Dispersing Agent on Ink Viscosity. *Nanomaterials* **2018**, *8*, 602. [CrossRef] [PubMed]
24. Lohse, D. Fundamental Fluid Dynamics Challenges in Inkjet Printing. *Annu. Rev. Fluid Mech.* **2022**, *54*, 349–382. [CrossRef]
25. Ivana, J.; Igor, K.; Ivana, T.; Dragoljub, N. PRINTING: Optical Paper Properties and Their Influence on Colour Reproduction and Perceived Print Quality. *Nord. Pulp Pap. Res. J.* **2013**, *28*, 264.
26. Peralba, M.; Dube, M.; Cormier, L.; Mangin, P. The Effect of Paper Properties on Print Quality. *Proc. TAGA* **1998**, *2007*, 22.
27. Adobe Express QR Code Creator. Available online: https://new.express.adobe.com/tools/generate-qr-code (accessed on 24 October 2023).
28. Suo, W.-J.; Goi, C.-L.; Goi, M.-T.; Sim, A.K.S. Factors Influencing Behavioural Intention to Adopt the QR-Code Payment: Extending UTAUT2 Model. *Int. J. Asian Bus. Inf. Manag.* **2022**, *13*, 1–22. [CrossRef]
29. Escobedo, P.; Ramos-Lorente, C.E.; Ejaz, A.; Erenas, M.M.; Martínez-Olmos, A.; Carvajal, M.A.; García-Núñez, C.; de Orbe-Payá, I.; Capitán-Vallvey, L.F.; Palma, A.J. QRsens: Dual-Purpose Quick Response Code with Built-in Colorimetric Sensors. *Sens. Actuators B Chem.* **2023**, *376*, 133001. [CrossRef]
30. Li, L.; Li, Y.; Wang, B.; Lu, J.; Zhang, S.; Yuan, W.; Wang, S.; Chang, C.C. A New Aesthetic QR Code Algorithm Based on Salient Region Detection and SPBVM. *Adv. Intell. Syst. Comput.* **2018**, *733*, 20–32. [CrossRef]

31. Peng, Y.; Zhang, L.; Song, Z.; Yan, J.; Li, X.; Li, Z. A QR Code Based Tracing Method for Fresh Pork Quality in Cold Chain. *J. Food Process Eng.* **2018**, *41*, e12685. [CrossRef]
32. Liang, K.; Chen, X.; He, R.; Li, J.; Okinda, C.; Han, D.; Shen, M. Development and Parameter Optimization of Automatic Separation and Identification Equipment for Grain Tracing Systems Based on Grain Tracers with QR Codes. *Comput. Electron. Agric.* **2019**, *162*, 709–718. [CrossRef]
33. Machado, J.G.D.C.F.; Nantes, J.F.D.; Leonelli, F.C.V. Do I Know What I Eat? The Use of QR Code in Food Packaging to Provide Traceability Information. *Int. J. Adv. Eng. Res. Sci.* **2019**, *6*, 45–58. [CrossRef]
34. Öblom, H.; Cornett, C.; Bøtker, J.; Frokjaer, S.; Hansen, H.; Rades, T.; Rantanen, J.; Genina, N. Data-Enriched Edible Pharmaceuticals (DEEP) of Medical Cannabis by Inkjet Printing. *Int. J. Pharm.* **2020**, *589*, 119866. [CrossRef]
35. Zhang, H.; Xu, Y.; Luo, M.; Zhang, C.; Gu, X.; Li, Z.; Ye, S. Integrated Food Quality Monitoring QR Code Labels with Simultaneous Temperature and Freshness Sensing in Real-Time. *J. Food Meas. Charact.* **2023**, *17*, 4834–4842. [CrossRef]
36. Żołek-Tryznowska, Z.; Bednarczyk, E.; Tryznowski, M.; Kobiela, T. A Comparative Investigation of the Surface Properties of Corn-Starch-Microfibrillated Cellulose Composite Films. *Materials* **2023**, *16*, 3320. [CrossRef] [PubMed]
37. *ISO 527-1*; Plastics Determination of Tensile Properties. Part 1: General Principles. ISO: Geneve, Switzerland, 2019.
38. Owens, D.K.; Wendt, R.C. Estimation of the Surface Free Energy of Polymers. *J. Appl. Polym. Sci.* **1969**, *13*, 1741–1747. [CrossRef]
39. *ISO 15989*; Plastics. Film and Sheeting Measurement of Water-Contact Angle of Corona-Treated Films. ISO: Geneve, Switzerland, 2004.
40. Leach, H.W. Structure of Starch Granule I. Swelling and Solubility Patterns of Various Starches. *J. Cereal Chem.* **1959**, *36*, 534–544.
41. Ahmed, J.; Thomas, L.; Arfat, Y.A.; Joseph, A. Rheological, Structural and Functional Properties of High-Pressure Treated Quinoa Starch in Dispersions. *Carbohydr. Polym.* **2018**, *197*, 649–657. [CrossRef]
42. Żołek-Tryznowska, Z.; Kałuża, A. The Influence of Starch Origin on the Properties of Starch Films: Packaging Performance. *Materials* **2021**, *14*, 1146. [CrossRef]
43. Dai, L.; Zhang, J.; Cheng, F. Effects of Starches from Different Botanical Sources and Modification Methods on Physicochemical Properties of Starch-Based Edible Films. *Int. J. Biol. Macromol.* **2019**, *132*, 897–905. [CrossRef]
44. Jagannath, J.H.; Nanjappa, C.; Das Gupta, D.K.; Bawa, A.S. Mechanical and Barrier Properties of Edible Starch-Protein-Based Films. *J. Appl. Polym. Sci.* **2003**, *88*, 64–71. [CrossRef]
45. Laohakunjit, N.; Noomhorm, A. Effect of Plasticizers on Mechanical and Barrier Properties of Rice Starch Film. *Starch/Stärke* **2004**, *56*, 348–356. [CrossRef]
46. *ASTM D523-14*; Standard Test Method For Specular Gloss. ASTM: West Conshohocken, PA, USA, 2018.
47. Izdebska, J. 1—Printing on Polymers: Theory and Practice. In *Printing on Polymers: Theory and Practice*; Izdebska, J., Thomas, S., Eds.; William Andrew Publishing: Amsterdam, The Netherlands, 2016; pp. 1–20, ISBN 978-0-323-37468-2.
48. Izdebska-Podsiadły, J. Effect of Plasma Surface Modification on Print Quality of Biodegradable PLA Films. *Appl. Sci.* **2021**, *11*, 8245. [CrossRef]
49. Meruga, J.M.; Cross, W.M.; Stanley May, P.; Luu, Q.; Crawford, G.A.; Kellar, J.J. Security Printing of Covert Quick Response Codes Using Upconverting Nanoparticle Inks. *Nanotechnology* **2012**, *23*, 395201. [CrossRef]
50. Tomašegović, T.; Pibernik, J.; Poljaček, S.M.; Madžar, A. Optimization of Flexographic Print Properties on Ecologically Favorable Paper Substrates. *J. Graph. Eng. Des.* **2021**, *12*, 37–43. [CrossRef]
51. Kim, B.-S.; Jung, H.-C.; Park, J.-Y. Investigation on Relationship between Pore Structure of Coating Layer and Ink Residual Behavior. *J. Korea TAPPI* **2002**, *34*, 53–58.
52. Ataeefard, M. Study of PLA Printability with Flexography Ink: Comparison with Common Packaging Polymer. *Prog. Color. Color. Coat.* **2019**, *12*, 101–105.
53. Cigula, T.; Tomašegović, T.; Hudika, T.; Donevski, D. Influence of the Ink and Substrate Properties on the Ink Transfer in Lithography. In Proceedings of the International Symposium on Graphic Engineering and Design, Belgrade, Serbia, 8–10 November 2018; pp. 45–50.

materials

MDPI

Article

Effect of Mass Reduction of 3D-Printed PLA on Load Transfer Capacity—A Circular Economy Perspective

Aneta Liber-Kneć [1,2] and Sylwia Łagan [1,*]

1 Faculty of Mechanical Engineering, Department of Applied Mechanics and Biomechanics, Cracow University of Technology, al. Jana Pawła II 37, 31-864 Kraków, Poland; aneta.liber@pk.edu.pl
2 Interdisciplinary Center for Circular Economy, Cracow University of Technology, ul. Warszawska 24, 31-155 Kraków, Poland
* Correspondence: sylwia.lagan@pk.edu.pl

Abstract: (1) Background: Optimizing infill density in 3D-printed PLA parts reduces material usage, cost, and waste. This study examines mechanical behavior in the initial and hydration stages. The findings provide valuable data for numerical simulations and engineering applications in additive manufacturing. (2) Methods: PLA specimens were printed with infill densities of 100%, 75%, and 25%. Mechanical tests, including tensile and compression tests, and one-hour stress-relaxation at 2% strain were conducted. The digital image correlation method was used to obtain the strain fields on the samples' surface under tensile loading. Mechanical properties, including the elastic modulus, strength values, and Poisson's ratio, were assessed. Hydrolytic degradation effects over one month were also evaluated. (3) Results: Lowering the PLA infill density reduced the ultimate tensile strength (from 60.04 ± 2.24 MPa to 26.24 ± 0.77 MPa), Young's modulus (from 2645.05 ± 204.15 MPa to 1245.41 ± 83.79 MPa), compressive strength (from 26.59 ± 0.80 MPa to 21.83 ± 1.01 MPa), and Poisson's ratio (from 0.32 to 0.30). A 40% mass reduction (form 100% to 25% infill density) resulted in a 56% decrease in tensile strength and a 53% decrease in Young's modulus. A 31% mass reduction was observed for compression samples. Stress relaxation decreased significantly from 100% to 75% density, with further reductions having minimal impact. Hydrated samples showed no mechanical changes compared to baseline specimens. (4) Conclusions: Optimizing infill density in 3D-printed PLA parts helps to balance mechanical performance with material efficiency. The best mechanical properties are typically achieved with an infill density of 100%, but results show that decreasing the mass of the part by a reduction in infill density from 75% to 25% does not significantly affect the ability to transfer tensile and compression loads. PLA's biodegradability makes it a viable alternative to stable polymers. By minimizing material waste and enabling the efficient use of resources, additive manufacturing aligns with the principles of a closed-loop economy, supporting sustainable development.

Keywords: polylactide; 3D printing; mechanical properties; digital image correlation; stress relaxation

1. Introduction

Plastics production exceeded the level of 400 million tons in 2023, and the trend of yearly production increases observed since 1950 suggests further growth [1]. Increasing production of plastic goods has led to the generation of around 350 million tons of plastic waste each year [2]. This shows the critical need to implement a circular plastic economy. The circular plastic economy promotes the flow of plastics in a closed cycle, in which plastic goods at the end of life will turn into resources, minimizing plastic pollution. A circular plastic economy can fall into several models: (1) reuse and extend service life through repair or remanufacture;

(2) recycling—turning old goods into new resources; (3) reducing the use of plastic or replacing petroleum-based materials with biobased and biodegradable polymers [3,4].

Substituting plastics derived from fossil resources with polymers from renewable resources brings several benefits related to the reduction of greenhouse gas emissions and obtaining polymers that are easier to recycle at the end of their life through the process of biodegradation [5]. The most commonly produced bioplastics are polylactic acid (PLA), polyhydroxyalkanoates (PHAs), starch, cellulose, and protein-based polymers [6]. Such polymers can be recycled through a degradation process in water, soil, and industrial composting systems, especially in cases where plastics become highly contaminated and are difficult to recover. Industrial composting systems meet the ISO 17088 [7] requirements for degradation (temperature, humidity, and microbial blends), the percentage of CO_2 emitted from bioplastics, and any toxic residues remaining. Among biodegradable polymers, polylactic acid draws the most attention. PLA is a biodegradable and biocompatible thermoplastic polyester, derived from renewable resources such as cornstarch and sugarcane. The use of natural raw materials means that PLA can degrade to carbon dioxide and water or lactide acid oligomers, reducing the environmental impact associated with plastic waste. The production of PLA through fermentation process in controlled polymerization of lactic acid monomers decrease carbon emission. Thus, it has become a promising alternative to conventional oil-based plastics [8,9]. In the context of a circular economy, PLA can be considered as a special polymer, because it can be treated at all levels of the waste management hierarchy including source reduction (reuse), recycling (mechanical or chemical), incineration with energy recovery, and composting at the end of its life [10–12].

The increased interest in PLA is due to its competitive prices, ease of processing, and biodegradability. These features also contribute to its industrial applications, which can be divided into two groups for economic reasons based on its useful life: long-lasting (useful life > 3 years) and short-lasting (useful life < 3 years) [10]. The PLA's properties and performance make it suitable for food packaging application, and in this aspect, it is widely used (except for carbonate beverage bottles) [13]. Another application area for PLA is the automotive sector, with the majority of PLA applications in automotive components, including items such as front panels, floor mats, car seats, pillar covers, door trims, and headliners [14]. Another field of application for biodegradable polymers, including PLA, is plasticulture, where the waste handling of conventional oil-based plastics carries risks of pesticide contamination and can limit their recycling. Thus, the use of biodegradable materials in this area is becoming increasingly popular [10]. PLA is also widely used in many different areas of medicine [10,15,16], such as orthopedics [17], dentistry [18,19], and ophthalmology [20]. Depending on its intended use, it can act as a temporary implant to support the healing process, e.g., as a cellular scaffold for bone reconstruction [21], as an attachment for ruptured ligaments or tendons [22], or in vascular stents [23], and, most importantly, in surgical sutures [24], which can biodegrade and be bioresorbable as the body takes over their function. The popularity of PLA in medicine is related not only to its biodegradability (decomposed into CO_2 and H_2O) and versatility of application but also to its biocompatibility (compatible with living tissues) and good strength properties [15]. PLA is also used in prosthetics as an alternative to the traditional composite socket in femoral stem prosthesis [25,26], and in upper limb orthoses and prostheses [27].

Despite the great promise of using PLA, there are still many challenges related to its unsatisfactory properties (e.g., mechanical, thermal) for given applications [28,29] and the design of cost-effective processing into intended forms and structures [10,30].

An equally important circular economy issue is the process of converting PLA into finished products. One of the methods of processing PLA is fused deposition modeling (FDM), an additive technology that can be used to create complex structures with great precision.

A circular plastic economy also addresses additive manufacturing as a technology used in biodegradable polymer processing [4] and can be an option for polymer recycling processes by using 3D-printing filaments derived from plastic waste [31,32]. The 3D printing of PLA is associated with many parameters that should be optimized to obtain a good performance of produced parts. Many studies show the influence of 3D-printing parameters on the mechanical behavior of parts made from PLA [4,33,34]. Authors have demonstrated that the mechanical performance of printed samples is influenced by infill density, infill pattern, raster angle, layer thickness, nozzle temperature, and bed temperature.

The results of the investigation conducted on 3D-printed PLA by Rodríguez-Panes et al. [33] showed that increasing the infill density improves mechanical strength, but a larger layer height had the opposite effect. A higher layer thickness reduces the strength of the component but, on the other hand, results in a shorter printing time. Orienting the layers perpendicular to the load direction had a negative effect on the tensile strength. Among the three-dimensional printing parameters analyzed, the authors identified filling density (the quantity of material with which the component is 3D-printed) as the key parameter. Albadrani compared the mechanical properties of 3D-printed PLA with 15° and 30° raster angles and reported that by selecting raster angles in the desired directions for printed objects, it is possible to enhance their strength and stiffness [35]. Alhazmi et al. [34] analyzed PLA printed dog bone specimens with three different raster orientations (±45°, 45°, and 0°) and infill densities (20%, 40%, 60%, 80%, and 100%). The optimal mechanical performance for PLA was achieved with a raster orientation of ±45°, showing higher values of tensile strength and Young's modulus in comparison to other orientations. Considerable variation in Young's modulus and tensile strength was observed at 40%, 60%, and 80% infill densities, showing the crucial role of this parameter. Kuclourya et al. [4] estimated that the tensile strength of 3D-printed PLA had a direct relationship with infill density. When moving from 30% infill to 60% infill, the tensile strength increased by 15%. It was also observed that at a 0°/90° raster angle orientation, the tensile strength was the highest, with a decrease at a ±45° raster angle.

Knowing the load capacity that 3D-printed PLA parts can transfer leads to a reduction of the infill density, and thus the amount of material used. This approach has several advantages during production, such as less material use and reduced manufacturing time. These factors influence cost reduction and decrease end-of-life waste. Therefore, in order to achieve a low weight and high strength of the product, it becomes necessary to obtain an optimal infill density. Recent studies on lightweight design in fused deposition modeling (FDM) have highlighted diverse strategies to reduce part mass without compromising structural integrity. Researchers have explored the role of variable infill densities in optimizing mechanical performance, while advances in biopolymer materials have introduced new sustainability dimensions to lightweight fabrication [4,33,34,36]. Additionally, topology optimization algorithms have become instrumental in distributing material more efficiently within printed components.

However, despite these developments, the literature insufficiently addresses the influence of mass reduction on load transfer behavior. The article concentrates on the influence of infill density on the mechanical properties of PLA printed specimens under tensile and compression load to indicate the optimal relationship between the mass of printed parts and the capacity for load transfer. This gap is critical, as changes in weight distribution can significantly alter how loads are transmitted through a part, potentially leading to unanticipated stress concentrations or failure modes. Further investigation is necessary to understand this relationship and ensure reliable design in weight-sensitive applications. An additional aspect was to carry out stress-relaxation tests, as well as to determine the Poisson's ratio using the digital image correlation method. This comprehensive approach to assessing the mechanical properties for different stress states provides data that can be useful for numerical analyses under different load states. The effect of hydration on mechanical properties was initially assessed by testing samples exposed to the liquid for one month. In our work, we analyze tensile and compressive loads for the samples with the same printing parameters, whereas other works focus on a single load scheme, which makes comparison of the results difficult.

2. Materials and Methods

2.1. Materials

The test specimens (for compression and tensile tests) were made using the FDM (fused deposition modeling) method. The 3D printer MarketBot SKETCH (MarketBot Industries, LLC, New York, NY, USA) was used with a working area of 150 × 150 × 150 mm and a layer resolution of 100–400 μm with UltiMaker Cura Cloud software V5.2 (UltiMaker, Zaltbommel, The Netherlands). The PLA Starter filament (ROSA PLAST SP. z o.o., Hipolitow, Poland), odorless, with a white color, a diameter of 1.75 mm, a density of 1.24 g/cm^3, and a heat distortion temperature of 55 °C, was used. The printing parameters are presented in Table 1.

Table 1. Basic printing parameters.

Printing Parameters	Value
speed of monolithic layers, mm/s	35
speed of non-monolithic layers, mm/s	50
working table temperature, °C	50
speed of extruder travel over areas where material was not applied, mm/s	80
thickness of the invisible layers of the upper part of the sample, mm	0.36
thickness of the outer top layer, mm	0.72
filament withdrawal distance, mm	5
printing temperature (print head), °C	220
infill type	Diamond Fill Fast
layer height, mm	0.2
number of contours	2
raster angle, °	0/90
density of infill, %	100, 75, 25

The Diamond Fill Fast pattern is a type of lattice with a geometric arrangement of diamonds. It features a repetitive layout of rhombuses or diamonds, which can be filled with various textures or left open, creating a mesh-like effect. Two types of specimens with infill densities of 100%, 75%, and 25% (Figure 1) were printed: dumbbell-shaped samples (Figure 1a) for tensile tests and two types of cylindrical specimens differing in their height-to-diameter ratio, i.e., 1.5× and 3× (Figure 1b) for compression tests. Tensile test samples were made in accordance with ISO 527-2 [37] with regard to cross-sectional dimensions and gauge length, and the width and length of the grip section of the specimen were reduced. The samples were printed in a single build orientation, known as the flat printing orientation, relative to the plane of the printer table in the Cartesian reference system, where XY determines the plane of the table and Z direction is perpendicular to the printer table. Each printed sample had two bottom/top layers with a thickness of 0.72 mm for outer layer and 0.36 mm for invisible layer. Every layer of infill pattern had the thickness of 0.2 mm.

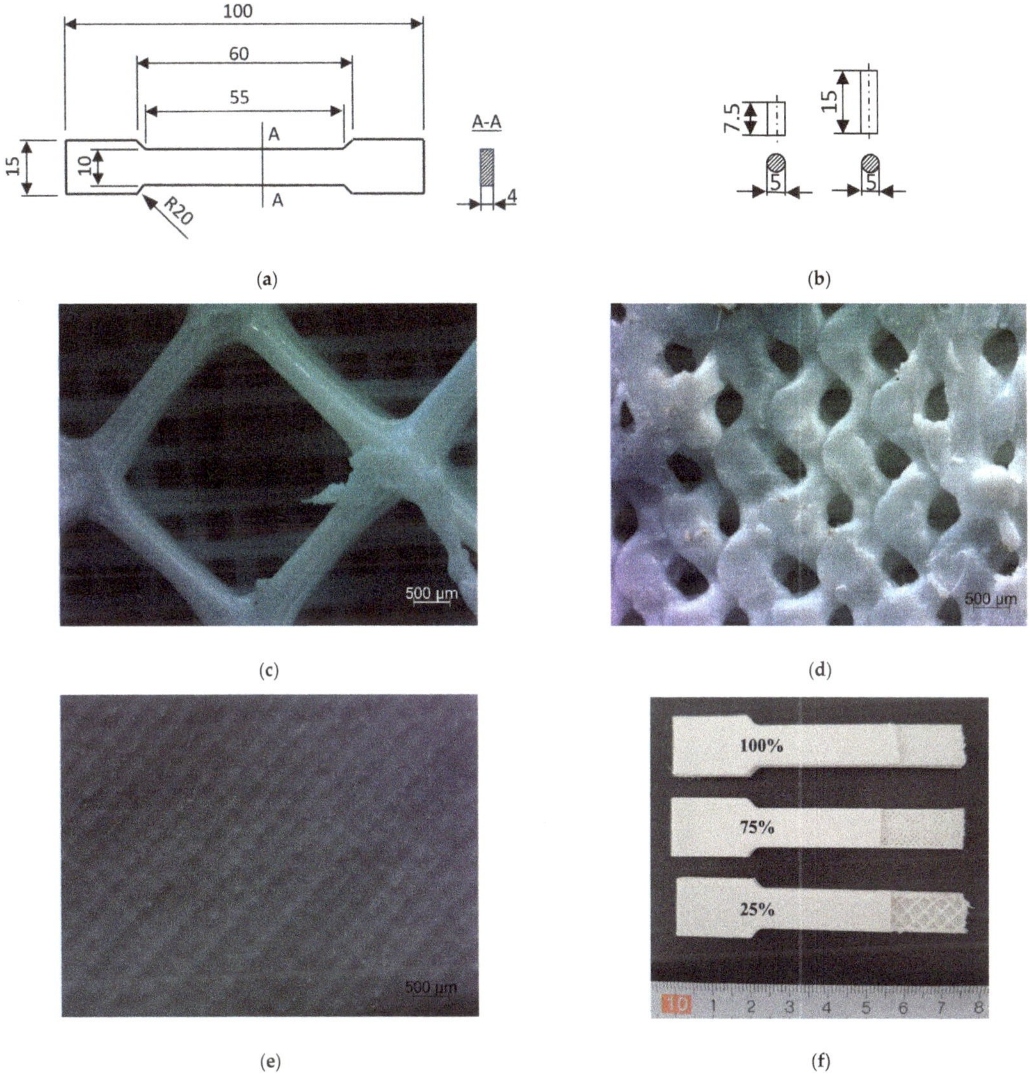

Figure 1. Printed specimens: (**a**) for tensile test, (**b**) for compression test, (**c**) internal structure of the sample with an infill density of 25%, (**d**) internal structure of the sample with an infill density of 75%, (**e**) internal structure of the sample with an infill density of 100%, (**f**) actual printed samples without outer layer.

2.2. Methods

2.2.1. Mechanical Tests

An MTS Insight 50 (MTS Insight^{TM}, Eden Prairie, MN, USA) universal testing machine with a ±50 kN load cell and the Test Works 4.0 software was used for tensile (in accordance with ISO 527 standard [37]) and compression tests. The compression speed was 0.5 mm/min. The specimens (n = 5) were compressed until their height was shortened by 2.0 mm. The tensile test speed was 5 mm/min and the gauge length was 60 mm. Samples (n = 5) were subjected to tension until destruction. One-hour stress-relaxation tests were also performed at 2% of tensile strain for PLA samples with three different infill densities (100%, 75%, and 25%).

All tests (compression, tensile, and stress relaxation) were carried out at a room temperature of 20 ± 1 °C and 40% humidity. After mechanical tests, the values of ultimate tensile strength (UTS) and compressive strength (σ_c) were determined and the values of elastic modulus (Young's modulus (E) and compressive modulus (E_c)). The tensile strength-to-weight ratio, and the tensile stiffness-to-weight ratio (both in MPa/g), of each specimen was determined by dividing the peak tensile stress (UTS*), and Young's modulus (E*) by the specimen mass, m (g), respectively. The value of residual stress (in MPa) after 1 h relaxation was obtained in initial and after incubation stage, and the normalized stress (-) curve were calculated by dividing the stress-relaxation data point by the peak tensile stress corresponding to 2% of stretching and mass of each specimen.

In addition, the effect of one-month hydrolytic degradation on the mechanical properties was considered.

2.2.2. Digital Image Correlation (DIC)

In order to assess the deformation capacity of the specimens across the entire surface during stretching and obtain values of Poisson's ratio, the method of digital image correlation (DIC) was used (Figure 2). The DIC method is an optical, non-contact measurement tool used to assess displacement and strain in three-dimensional space. The DIC system (Dantec Dynamics GmbH, Ulm, Germany) consisted of two 2 MPx cameras and the DIC data processing system Istra4D V4.8.2.248 (Dantec Dynamics GmbH, Ulm, Germany). All samples were covered with a speckle pattern (base surface—matte white acrylic paint, dots—spraying matte black acrylic paint). Calibration was performed before the tests using a proprietary calibration target (Al11-BMB-9 × 9, Dantec Dynamics GmbH, Ulm, Germany). The size of the camera's field of view was limited to the region of interest (ROI) (about 80% of sample surface) to avoid edge effects. During the tensile test, the cameras tracked the movement of the speckle pattern and then during the analysis, the recorded images were transformed into a 3D surface mesh using stereo-triangulation. By correlating the reference surface mesh (for the unloaded specimen) and the subsequent generated meshes, a map of the displacement and deformation on the specimen surface in three directions was obtained. The Poisson's ratio was calculated and the base of strain values for longitudinal and transversal directions to force direction.

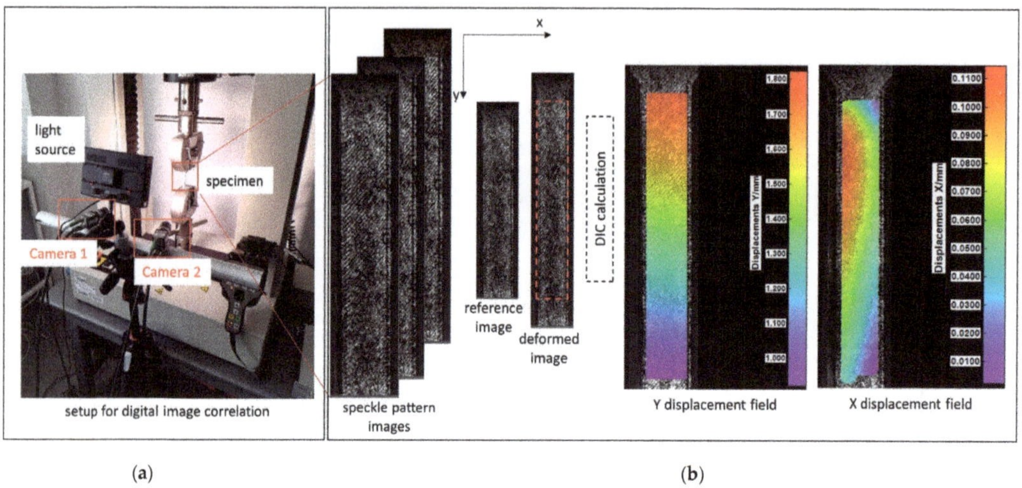

Figure 2. Two-dimensional digital image correlation: (**a**) system configuration, (**b**) in-plane displacement field calculation.

2.2.3. Hydration Test

From the group of specimens of each infill density intended for tensile and compression tests (15 mm height), five specimens were placed in Ringer's solution (Serumwerk Bernburg AG, Benburg, Germany): sodium chloride 8.6 g, potassium chloride 0.3 g, calcium chloride dihydrate 0.33 g 96 (^mmol/L: Na^+ 147.2, K^+ 4, Ca^{++} 2.25, Cl^- 155.7), pH value 5.0–7.5, acidity (titration to pH 7.4) <0.1 mmol/L, theoretical osmolarity 307 mOsm/L at 37.0 ± 0.1 °C in a thermal chamber (AL01-02-100 Advantage Lab, Schilde, Belgium) for 30 days. The volume of the solution was 100 mL for the compression tests specimens (cylindrical shape) and 500 mL for the tensile tests specimens (dumbbell-shaped). The mass of the samples was controlled using an analytical balance (AS 160/C/2, Radwag, Radom, Poland) with an accuracy of 1×10^{-4} g, before and after the incubation period. The degree of mass change was assessed as the quotient of the post-incubation mass change (the difference between the initial and post-incubation mass) and the initial weight. The initial mass of the dumbbell-shaped specimens was 4.2979 ± 0.7499 g, 3.9889 ± 0.6294 g, and 3.6786 ± 0.0862 g, and that of the 15 mm high cylindrical specimens was 0.2339 ± 0.042 g, 0.2087 ± 0.0023 g, and 1.685 ± 0.0059 g for infill densities of 100%, 75%, and 25%, respectively. After the 30-day incubation process, the degree of absorption was determined.

2.2.4. Statistical Analysis

The data presented in this study were the mean values ± standard deviation determined by testing five samples for each case considered. Significant statistical differences between groups of data were determined using Student's *t*-test. The level of significance used was 95 per cent (i.e., $p \leq 0.05$). The basic assumptions of Student's *t*-test were checked, i.e., equality of groups (same number of observations) and homogeneity of variance (Fisher's test).

3. Results

The change in mass of dumbbell-shaped samples and cylindrical samples after 30 days of hydration is shown in Figure 3. Evaluation of the effect of the short-term hydration of the samples showed that, regardless of the level of filling, the samples showed stability and their weight changes were not significant. This was confirmed in the mechanical properties tests, where the tested parameters for hydrated samples did not change compared to the base samples.

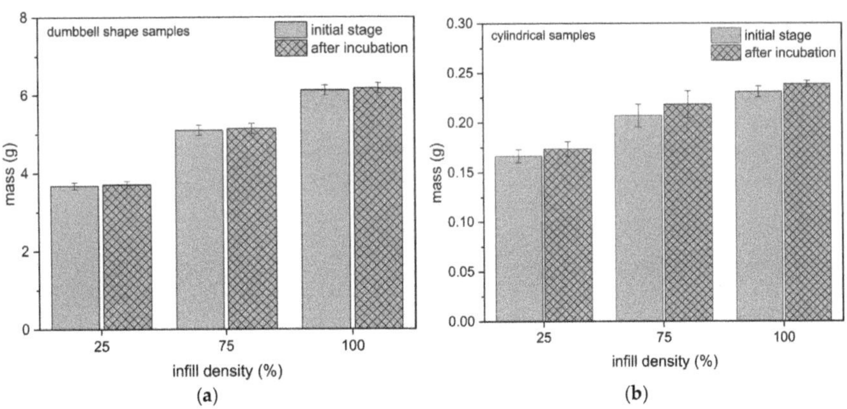

Figure 3. The mass change for (**a**) dumbbell-shaped samples, (**b**) cylindrical samples (3× ratio).

The use of different levels of infill density resulted in a mass redaction of 40% between samples with a 100% and 25% infill density (for samples used for tensile and relaxation tests), and for cylindrical samples a reduction of 28% of the mass for triplicate samples and 33% for 1.5× samples.

3.1. Tensile and Compression Tests

During the tensile test, all samples (regardless of infill density) were characterized by a nonlinear response of the material to loading (Figure 4). The highest values of the breaking force were obtained for samples in their initial state with 100% filling (2401.4 ± 89.7 N). Samples with a 75% infill of density obtained 1386.3 ± 63.7 N, while the group with the lowest infilling density obtained 1049.8 ± 30.9 N. The recorded extension (corresponding to the maximum force) was at the level of 2.5 ± 0.2 mm, 2.2 ± 0.2 mm, and 2.0 ± 0.1 mm for samples with a 100%, 75%, and 25% density of infill, respectively. In each group (100%, 75%, and 25%), the 30-day incubation caused a decrease in the value of the maximum force by 8%, 5%, and 1%, respectively, and in the corresponding elongation by 32%, 27%, and 20%, respectively. The infill density of the specimen had a significant effect on the tensile curve, whereas the effect of hydration was only significant for specimens with a 100% infill.

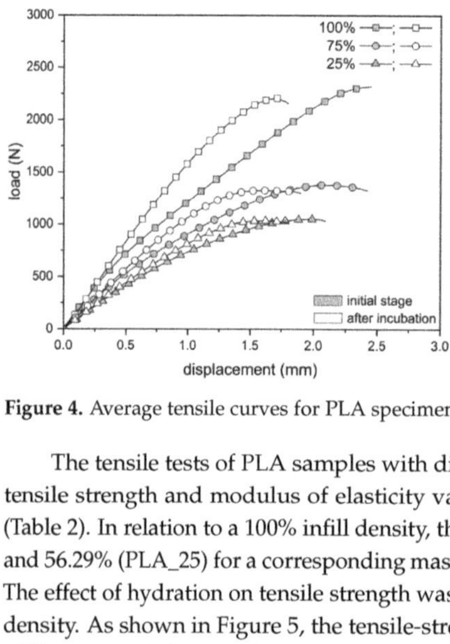

Figure 4. Average tensile curves for PLA specimens.

The tensile tests of PLA samples with different infill densities showed a decrease in tensile strength and modulus of elasticity values along with a decrease in infill density (Table 2). In relation to a 100% infill density, the UTS values decreased by 42.27% (PLA_75) and 56.29% (PLA_25) for a corresponding mass reduction of 7.18% and 14.41%, respectively. The effect of hydration on tensile strength was statistically significant only for a 100% infill density. As shown in Figure 5, the tensile-strength-to-weight ratio (UTS*) and the tensile stiffness-to-weight ratio (E*) relating the obtained parameters to the mass of the samples resulted in a significant decrease in their values when the filling density was reduced to 75%. A further reduction in infill density from 75% to 25% did not significantly reduce the analyzed values. The values of specific strength were 9.8 ± 0.3 MPa/g, 6.8 ± 0.3 MPa/g, and 7.4 ± 0.3 MPa/g for the 100%, 75%, and 25% filling densities, respectively.

Table 2. Mechanical properties obtained in tensile test.

	UTS [MPa]	E [MPa]
PLA_100	60.04 ± 2.24 ^{a,b}	2645.05 ± 204.15 ^{a}
PLA_100i	55.50 ± 1.16 ^{b}	2693.63 ± 123.01
PLA_75	34.66 ± 1.59 ^{a}	1635.93 ± 149.27 ^{a}
PLA_75i	33.17 ± 1.20	1756.61 ± 87.87
PLA_25	26.24 ± 0.77 ^{a}	1245.41 ± 83.79 ^{a}
PLA_25i	25.98 ± 0.47	1329.15 ± 97.23

Note: PLA_100/PLA_75/PLA_25–100%/75%/25% of infill density, i—specimens after hydration. Note: The same lowercase letters within the columns indicate a statistically significant difference between the groups ($p \leq 0.05$).

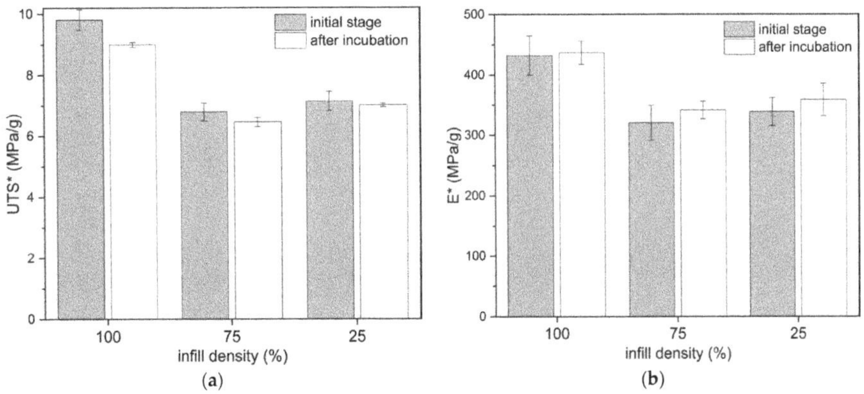

Figure 5. Comparison of the influence of infill density of PLA dumbbell-shaped specimens on the tensile properties in relation to the mass: (**a**) ultimate tensile strength (UTS*), (**b**) Young's modulus (E*).

Analyzing the shapes of the average compression curves (Figure 6), linear and nonlinear regions can be seen. The test specimens were characterized by linear stiffness up to a force level of about 300 N and then deformed plastically. The infill density clearly affects the material's response during compression, whereas the load exceeded the yield strength.

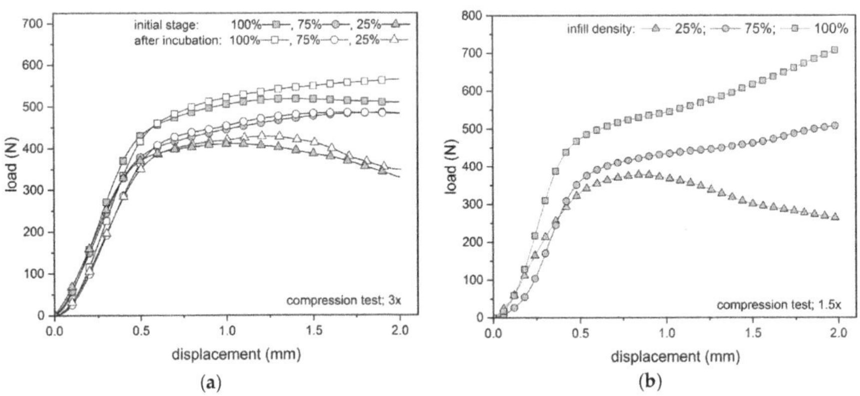

Figure 6. Comparison of the influence of infill density of PLA cylindrical specimens on the compressive characteristics: (**a**) average compressive curves before and post incubation (3× ratio samples), (**b**) average compressive curves of 1.5× ratio samples.

The character of fracture under tensile strain was brittle without delamination between the layers of printed samples (Figure 7). In contrast, during compression, regardless of the multiplicity of the specimens, the plastic nature of the failure was observed (the specimens assumed a barrel shape) with no signs of cracks between the layers. This is also confirmed by the shape of the compression curve.

By evaluating the results of the compression tests (Table 3), it can be noted that in the group of 1.5× specimens, the difference between values of compressive modulus was not statistically significant ($p \geq 0.05$). Similarly, in the group of 3× specimens, the differences in modulus values were not statistically significant. The one-month incubation of the samples did not significantly affect the variation of the compressive modulus values. Reducing the filling density of the samples, both 1.5× and 3×, resulted in a reduction in the compressive strength values (statistically significant differences between the groups). The reduction in the UTS also depended on the height of the samples used in the tests. For 1.5× samples, the UTS varied from 28.42% (PLA_75) to 46,42% (PLA_25), and for 3× samples, it varied from 6.39% (PLA_75) to 17.90% (PLA_25) for a corresponding mass reduction of 7.22% and 14.45%, respectively. This showed the significant influence of the ratio of specimen height to diameter during the compression test on the results obtained. The effect of incubation on the compressive strength was statistically significant ($p \leq 0.05$) only for specimens with 100% fill density.

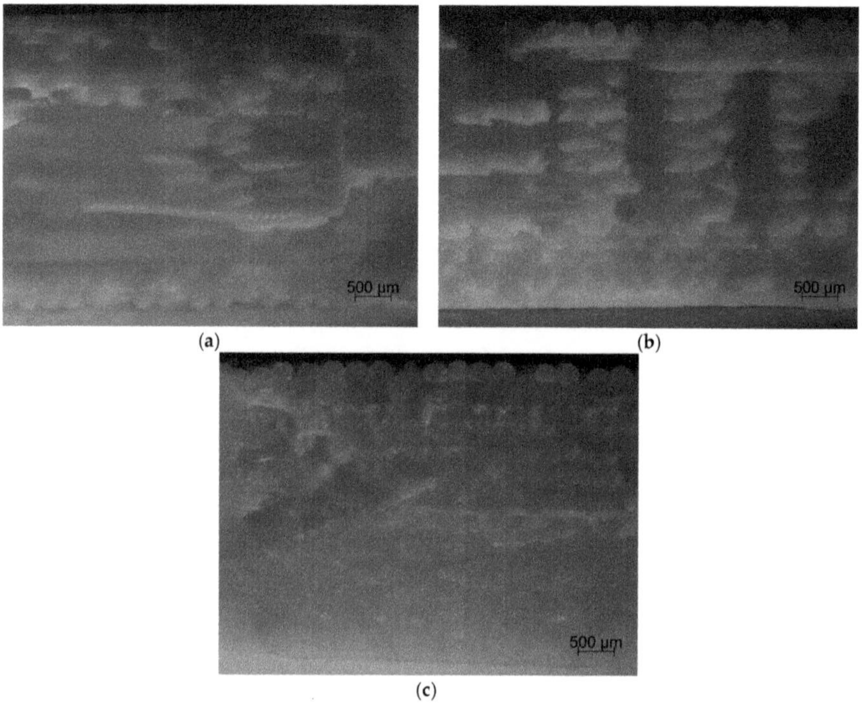

Figure 7. Comparison fracture of the specimen view: (**a**) fracture of the sample with an infill density of 25%, (**b**) fracture of the sample with an infill density of 75%, (**c**) fracture of the samples with an infill density 100%.

Table 3. Mechanical properties obtained in compression tests.

	σ_c [MPa]	E_c [MPa]
1.5_PLA_100	36.24 ± 3.32 [a,d]	504.60 ± 9.20
1.5_PLA_75	25.94 ± 1.82 [a]	513.60 ± 11.91
1.5_PLA_25	19.42 ± 0.44 [a,e]	516.60 ± 7.12
3_PLA_100	26.59 ± 0.80 [b,c,d]	513.2 ± 9.68
3_PLA_100i	28.92 ± 0.96 [c]	513.2 ± 4.87
3_PLA_75	24.89 ± 0.88 [b]	515.00 ± 7.01
3_PLA_75i	25.32 ± 1.62	515.00 ± 8.74
3_PLA_25	21.83 ± 1.01 [b,e]	514.60 ± 3.61
3_PLA_25i	22.22 ± 1.24	517.40 ± 3.14

Note: PLA_100/PLA_75/PLA_25–100%/75%/25% of infill density, i—specimens after hydration, 1.5/3—height to diameter ratio. Note: The same lowercase letters within the columns indicate a statistically significant difference between the groups ($p \leq 0.05$).

Relating the value of Young's modulus to the mass of the specimens tested showed that a reduction in the infill density did not reduce the compressive stiffness of the specimen (Figure 8a,c). For tensile strength, a decrease in tensile modulus was observed when the infill density was reduced from 100% to 75% (Figure 8b,d). A further reduction in infill density to 25% did not reduce the compressive strength in relation to mass.

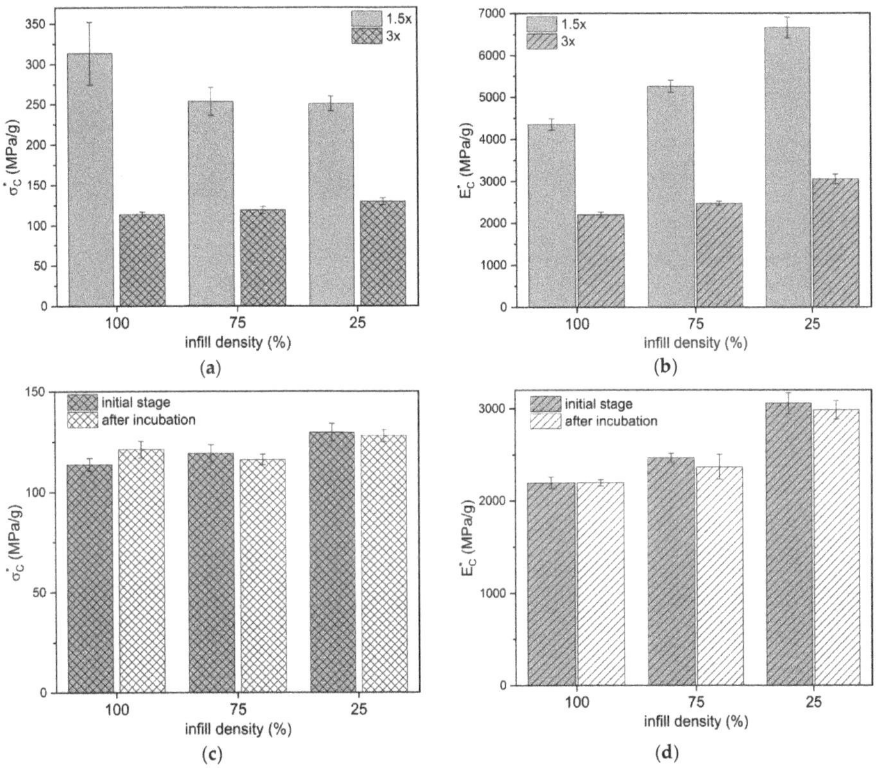

Figure 8. Comparison of the influence of infill density of PLA cylindrical specimens on the compressive properties in relation to mass: (**a**) compressive strength, (**b**) compressive modulus, (**c**) compressive strength of 1.5× ratio specimens, (**d**) compressive modulus before and post incubation (3× ratio samples).

3.2. Digital Image Correlation (DIC)

Deformations in the longitudinal (Y) and transverse (X) directions were recorded by the digital image correlation (DIC) system. The results of the uniaxial tension tests coupled with their corresponding DIC tensile displacement contours at frames captured just before fracture are presented in Figures 9–11. Based on the displacement maps, the strains for the two main directions were determined and the Poisson's ratio values were calculated (Figure 12). Determination of the Poisson's ratio using a digital image correlation system is possible through the use of a virtual gauge element. This analysis uses a virtual gauge element in the form of a polygon, within a predefined region of interest (ROI)—the area in which the Istra4D software V4.8.2.248 collects and averages the principal strains. Reducing the infill density of the PLA samples resulted in the same decrease in the Poisson's ratio value for both infill densities, i.e., from 0.32 to 0.30 (Table 4).

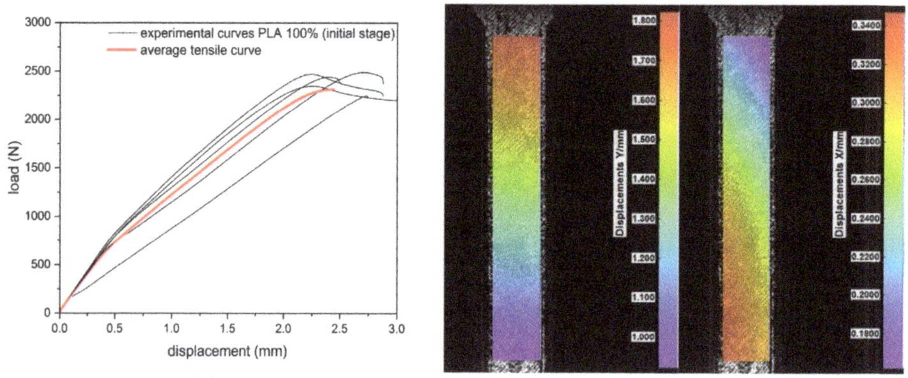

(a) (b)

Figure 9. Results of tensile test: (**a**) tensile curves for specimens of 100% infill density, (**b**) digital image correlation displacement distribution maps in the longitudinal (Y) and transverse (X) directions.

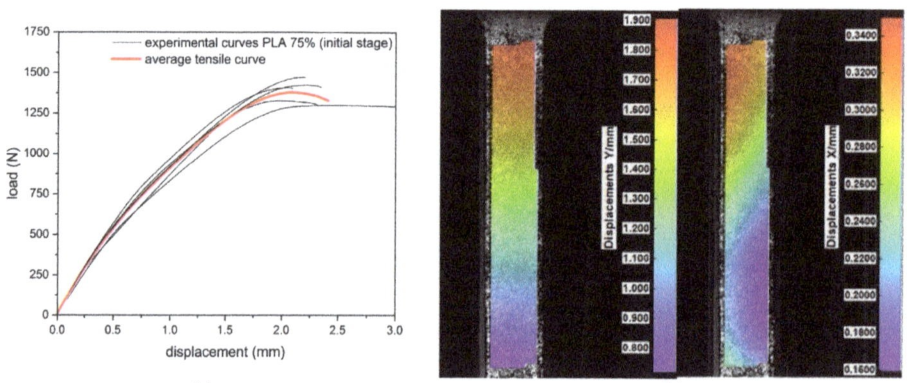

(a) (b)

Figure 10. Results of tensile test: (**a**) tensile curves for specimens of 75% infill density, (**b**) digital image correlation displacement distribution maps in the longitudinal (Y) and transverse (X) directions.

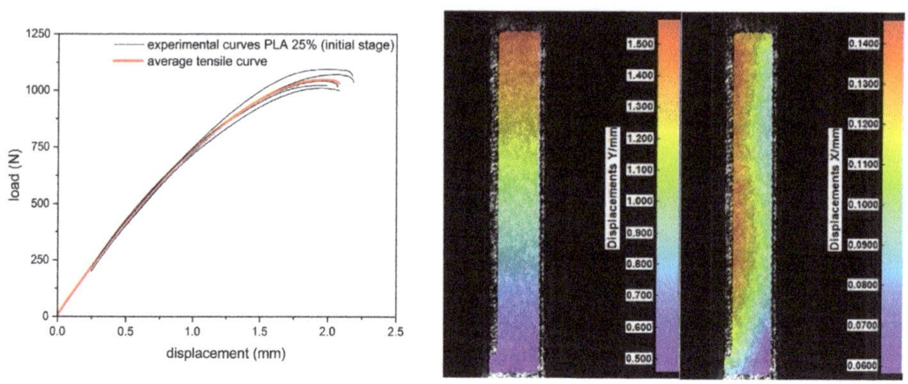

(a) (b)

Figure 11. Results of tensile test: (**a**) tensile curves for specimens of 25% infill density, (**b**) digital image correlation displacement distribution maps in the longitudinal (Y) and transverse (X) directions.

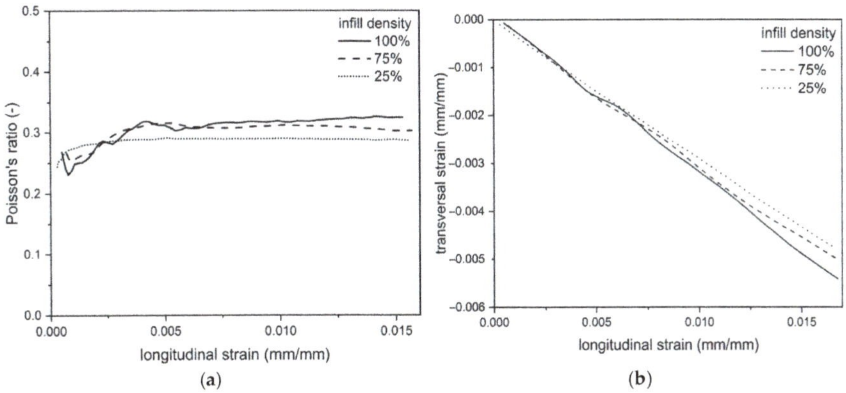

Figure 12. The curves obtained from DIC method for PLA sample materials with different infill densities: (**a**) Poisson's ratio vs. longitudinal strain, (**b**) transverse strain vs. longitudinal strain.

Table 4. Comparison of Poisson's ratio for PLA samples determined by DIC.

	Poisson's Ratio [-]
PLA_100	0.32 ± 0.01 [a,b]
PLA_75	0.30 ± 0.01 [a]
PLA_25	0.30 ± 0.01 [b]

Note: PLA_100/PLA_75/PLA_25–100%/75%/25% of infill density. Note: The same lowercase letters within the columns indicate a statistically significant difference between the groups ($p \leq 0.05$).

3.3. Stress Relaxation

The trend in stress relaxation was similar for all the specimens tested, with rapid relaxation in the first 100 s, followed by a slow slope to a stabilized stress value (Figure 13). Preliminary tests of stress relaxation at 2% strain highlighted differences between samples with 100% infill density and those with 75 and 25% infill (Table 5). As with tensile and compressive loading, a noticeable decrease in stress relative to the mass of the specimens is observed when the infill density is reduced from 100% to 75%, whereas further reductions in density do not result in a decrease in stress values.

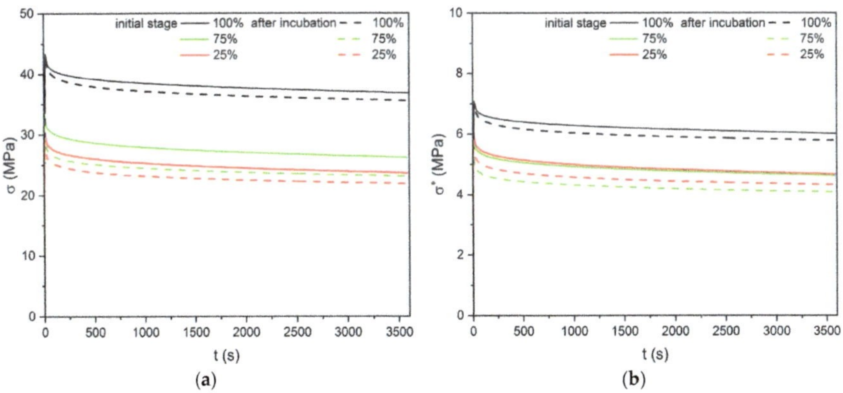

Figure 13. Stress-relaxation curves for PLA with different infill density: (**a**) stress relaxation vs. time, (**b**) stress relaxation related to mass vs. time.

Table 5. Values of maximal σ_{max}, residual σ_{∞}, and normalized $\sigma_{normalized}$ stress relaxation for PLA samples with different infill densities.

	σ_{max} [MPa]	σ_{∞} [MPa]	σ^*_{max} [MPa/g]	σ^*_{∞} [MPa/g]	$\sigma_{normalized}$ [-]
PLA_100	43.53 ± 1.54 [a,b]	36.80 ± 3.33 [a,b]	7.11 ± 0.25	6.02 ± 0.28	0.85 ± 0.01
PLA_100i	43.16 ± 0.40	35.55 ± 0.29	7.03 ± 0.03	5.79 ± 0.02	0.82 ± 0.00
PLA_75	33.60 ± 1.31 [a]	26.26 ± 1.10 [a]	5.93 ± 0.23	4.64 ± 0.20	0.78 ± 0.00
PLA_75i	30.09 ± 1.16	23.20 ± 1.32	5.32 ± 0.21	4.10 ± 0.24	0.78 ± 0.08
PLA_25	30.62 ± 0.76 [b]	23.72 ± 1.02 [b]	6.04 ± 0.15	4.83 ± 0.05	0.78 ± 0.02
PLA_25i	28.11 ± 0.90	21.94 ± 0.04	5.55 ± 0.18	4.33 ± 0.01	0.79 ± 0.03

Note: PLA_100/PLA_75/PLA_25–100%/75%/25% of infill density. i—specimens after hydration. Note: Same lowercase letters within the columns indicate statistically significant difference between the groups ($p \leq 0.05$).

4. Discussion

Reports on the influence of infill density on the mechanical properties of 3D-printed PLA are thoroughly documented by several authors [38–40]. The mechanical properties under tensile loading were of main interest in these publications. The result of this study concentrated on a more complex approach and provided data for tensile and compressive behavior, as well as stress relaxation under constant strain in PLA. Decreasing the infill density of 3D-printed PLA specimens resulted in a reduction in mass but also mechanical properties under tensile loading, including tensile strength and Young's modulus, indicating deteriorated stiffness. Significantly, a comparison of these parameters to the weight of the specimens showed that a reduction in infill density from 75% to 25% did not significantly reduce them. In the case of compression specimens, only the strength decreased with a reduction of infill density, whereas the values of the elastic modulus did not change significantly between the test groups. While tensile testing data is fairly common in the literature for 3D-printed specimens, the current work presented here reveals full-field longitudinal and transverse displacement contour plots highlighting the different displacement patterns observed for the various infill densities. DIC was also used to measure Poisson's ratio. The DIC method complements traditional testing approaches by offering a more comprehensive view of material behavior under stress.

Pandzic and Hodzic [41] indicated that the infill pattern has an influence on the tensile mechanical properties (tensile strength and elastic modulus), and they referred to values of ultimate tensile strength in the range from 38.40 to 45.57 MPa and Young's modulus in the range from 2.10 to 2.4 GPa, for 60% infill density. In another work [38], the authors considered the density of the infill and its gradient, and also noted the dependence of the strength properties on these parameters. The compressive strength increased with the increase in the degree of filling. Considering the change in the printing density of cylindrical specimens in the range of 80% to 10%, they recorded compressive strength values ranging from 41.2 MPa to 0.8 MPa and modulus of elasticity values from 1500 MPa to 10 MPa, respectively. On the other hand, by isolating two printing zones (the outer one at a constant density level of 80% and the inner one varying from 60% to 20%), they recorded a decrease in compressive strength from 31.7 MPa to 10.7 MPa, and a decrease in modulus of elasticity values from 1200 MPa to 500 MPa, respectively [38]. Dave H. et al. compared the effects of three infill density levels of 80%, 70%, and 60% and three layer heights (0.1, 0.2, and 0.3 mm) on compressive strength, obtaining values in the range of 33.13–52.93 MPa, 34.85–43.62, and 26.46–33.54, respectively [39]. Hodzic D. and Pandzic A. [40] emphasized that the mechanical properties of 3D-printed PLA material depend heavily on the type of infill pattern. They compared different infill patterns—concentric, grid, gyroid octet, and triangle—as well as their densities (20%, 40%, 60%, and 80%). Their findings indicate that the grid pattern at an 80% infill density exhibited the highest compressive strength and modulus values. Specifically, for 20% infill, they recorded 29.1 MPa and 0.49 GPa; for 40%,

38 MPa and 0.54 GPa; for 60%, 44.6 MPa and 0.67 GPa; and for 80%, 54.1 MPa and 0.66 GPa. The tensile test described in the paper [42] examines 3D-printed PLA material with a infill density of 30%, comparing various infill patterns—Hilbert, gyroid, 3D honeycomb, stars, and honeycomb. The results indicate that the honeycomb infill exhibits the highest tensile strength, reaching 29.4 MPa.

The ratio of the printed part's density to the bulk material's density is directly proportional to the change in Poisson's ratio, especially in orthotropic (directionally dependent) materials. Poisson's ratio in 3D-printed materials is highly tunable and depends on a combination of material selection, microstructural design, print path, orientation, and density. Print direction and raster orientation can introduce anisotropy, making Poisson's ratio vary depending on the direction of loading relative to the print layers [43]. The measured Poisson's ratio value for 100% infill density of 0.32 is in good agreement with the data provided by Wang et al. [44] in the range of 0.32 to 0.33 depending on the raster orientation and reported in the paper [45] for printed PLA at 0.33. The slight differences in the variation of Poisson's ratio values observed in our study, as well as in the literature, may be due to the dominant influence of the outer layers on the printed samples' deformation in the longitudinal and transversal directions compared to the core deformation for the given density and infill pattern.

As the results from the literature analysis show, the infill density [4,38,39], type of infill pattern [41,42], and the direction of 3D processing [34,35,46,47] of the specimens influence their mechanical behavior. The processing (printing) direction leads to significant anisotropy in properties such as strength, stiffness, ductility, and failure modes. Specimens printed with layers aligned along the loading direction show much higher strength and stiffness compared to those printed upright or at higher angles (e.g., 90°) [48]. Tüfekci et al. showed that the printing direction has a major impact on the tensile strength. The obtained tensile strength values were at the level of 42, 36, and 25 MPa for PLA samples with 100% infill density and 0°, 45°, and 90° infill orientations, respectively [46], whereas the stress-relaxation results presented in [46] focused on the change in initial stress values from 16.33, 13.48, and 9.55 MPa, for the 0°, 45°, and 90° infill orientations to 13. 89, 11.81, and 8.21 MPa, after two hours of relaxation.

The stress relaxation of PLA material was the subject of the paper [47], in which the authors referred to Young's modulus values of 3045 ± 3 MPa, 2914 ± 3 MPa, and 2932 ± 3 MPa for a 0°, 45°, and 90° infill orientation and normalized creep modulus values in the range of 0.85–0.89, 0.9–0.92, and 0.85–0.92 after about 10 min of relaxation for a 0°, 45°, and 90° infill orientation, respectively.

Systemic evaluation of how 3D-printed PLA performs under different load conditions can contribute to optimizing processing parameters, ensure material quality, and enable the reuse of PLA in new products, thus reducing waste and resource consumption. The results of this study can be used to evaluate how recycled and reprocessed PLA perform. Repeated PLA recycling can lead to decreases in tensile strength (from 20.58 MPa for virgin to 12.04 MPa for 10% virgin/90% recycled) and Young's modulus (from 5.86 GPa for virgin to 4.74 GPa for 10% virgin/90% recycled) [49]. Some studies report that adding recycled PLA to virgin PLA can improve its mechanical properties, from 44.2 ± 2.18 MPa for virgin PLA to 52.61 ± 2.28 MPa for 25% virgin/75% recycled PLA [50].

Overall, increasing the infill density of 3D-printed PLA components enhances their mechanical properties, including tensile strength, Young's modulus, fatigue resistance, and impact strength [51–53]. These improvements make high-density infill settings preferable for applications requiring robust and durable components. However, higher infill densities result in increased mass because more material is used to fill the internal structure of the printed object. The experimental data obtained in this study can be used in the design of 3D-printed components with consideration of different infilling densities in order to optimize the mass and mechanical properties of the printed components.

5. Conclusions

The closed-loop economy, which aims to improve sustainability, is turning the attention of manufacturers toward the thoughtful use of plastics throughout the value chain, forcing a different approach to product design and manufacturing. One such path may be to reduce the weight of the product. It may lower mechanical performance but can still provide environmental benefits by reducing the volume of plastics on the market and consequently plastic waste. The results obtained in this study show that reducing the mass of the samples by reducing the infill density from 100% to 75% reduces mechanical properties, but a further reduction to 25% does not significantly decrease the strength properties. When decreasing the infill density from 100% to 25%, a 40% mass reduction was observed, which resulted in a 56% decrease in tensile strength and a 53% decrease in Young's modulus. A 46% and 20% decrease in compressive strength was also observed, depending on the sample multiplication factor (1.5× and 3×, respectively). This shows that adjusting the infill density and appropriate design strategy for 3D-printed PLA can minimize material use while maintaining mechanical strength. This provides sustainability benefits by reducing material consumption, energy use, and production time. Using less material aligns with circular economy principles and supports the use of recycled or biodegradable feedstocks, further reducing environmental impact. PLA is biodegradable and, when sourced sustainably, has a lower carbon footprint than petroleum-based plastics. Incorporating recycled PLA in future studies can further reduce environmental impact. Increasing the infill density of 3D-printed PLA parts results in a heavier object with enhanced mechanical properties, including tensile strength, stiffness, and fatigue resistance. However, the optimal infill density may vary depending on the specific application requirements, balancing material use and the desired mechanical performance. In addition, the ability of PLA to slowly biodegrade is an advantage in favor of its use as a substitute for many stable polymers, the use of which is temporarily defined or limited. Lightweight components manufactured using 3D-printing technology are used as non-load-bearing structures in many areas of engineering, e.g., as housings and covers, prototypes, and conceptual and educational models, as well as everyday or decorative elements.

However, the methods of assessing the material parameters sought by the manufacturer should always be analyzed. In this work, attention was paid to the influence of both the types of tests and the shapes of the samples used in the tests, as well as the level of hydration and mass reduction, on the mechanical properties of PLA samples obtained by 3D printing. The data can be used for numerical analyses of objects manufactured using 3D printing with the application of various parameters, which can speed up the product optimization process. Using fused deposition modeling can support the transition towards a circular economy in the construction industry, particularly in medicine, which requires product personalization. The study has several limitations such as focusing on a limited set of printing parameters, limited time of water absorption, and lack of thermal and photo aging. The search for the load transfer capacity of 3D-printed PLA should be completed by testing under dynamic load. Future work on assessing the effect of mass reduction will focus on obtaining an optimal set of 3D-printing parameters to increase the ability to carry static and dynamic loads. Mechanical properties' evaluation will be enhanced by analysis of how the infill density, infill pattern, and raster orientation affect stress concentration, fracture resistance, and the overall mechanical performance of 3D-printed parts.

Author Contributions: Conceptualization, A.L.-K. and S.Ł.; methodology, S.Ł. and A.L.-K.; software A.L.-K. and S.Ł.; validation A.L.-K. and S.Ł.; formal analysis, A.L.-K. and S.Ł.; investigation, A.L.-K. and

S.Ł.; resources, A.L.-K. and S.Ł.; data curation, S.Ł. and A.L.-K.; writing—original draft preparation, A.L.-K. and S.Ł.; writing—review and editing, A.L.-K. and S.Ł.; visualization, S.Ł. and A.L.-K.; supervision, A.L.-K. and S.Ł.; project administration, A.L.-K.; funding acquisition, A.L.-K. and S.Ł. All authors have read and agreed to the published version of the manuscript.

Funding: This research received no external funding.

Institutional Review Board Statement: Not applicable.

Informed Consent Statement: Not applicable.

Data Availability Statement: The original contributions presented in this study are included in the article. Further inquiries can be directed to the corresponding author.

Conflicts of Interest: The authors declare no conflicts of interest.

References

1. Global Plastic Production 1950–2023, Published by Statista Research Department. Available online: https://www.statista.com/ (accessed on 8 May 2025).
2. Plastic Waste Worldwide—Statistics & Facts, Published by Bruna Alves. Available online: https://www.statista.com/ (accessed on 8 May 2025).
3. King, S.; Locock, K.E.S. A circular economy framework for plastics: A semi-systematic review. *J. Clean. Prod.* **2022**, *364*, 132503. [CrossRef]
4. Kuclourya, T.; Monroy, R.; Ahmad, R. Design of experiments to compare the reprocessing effect with Fused Deposition Modeling printing parameters on mechanical properties of polylactic acid specimens towards circular economy. *Prog. Rubber Plast. Recycl. Technol.* **2023**, *39*, 111–140. [CrossRef]
5. Sheldon, R.; Norton, M. Green chemistry and the plastic pollution challenge: Towards a circular economy. *Green Chem.* **2020**, *22*, 6310–6322. [CrossRef]
6. Karan, H.; Funk, C.; Grabert, M.; Oey, M.; Hankamer, B. Green bioplastics as part of a circular bioeconomy. *Plant Sci. J.* **2019**, *24*, 237–249. [CrossRef]
7. *ISO 17088:2021*; Plastics—Organic Recycling—Specifications for Compostable Plastics. ISO: Geneva, Switzerland, 2021.
8. Mosomi, E.K.; Olanrewaju, O.A.; Adeosun, S.O. Pivotal role of polylactide in carbon emission reduction: A comprehensive review. *Eng. Rep.* **2024**, *6*, e12909. [CrossRef]
9. Yu, J.; Xu, S.; Liu, B.; Wang, H.; Qiao, F.; Ren, X.; Wei, Q. PLA bioplastic production: From monomer to the polymer. *Eur. Polym. J.* **2023**, *193*, 112076. [CrossRef]
10. Castro-Aguirre, E.; Iñiguez-Franco, F.; Samsudin, H.; Fang, X.; Auras, R. Poly(lactic acid)—Mass production, processing, industrial applications, and end of life. *Adv. Drug Deliv. Rev.* **2016**, *107*, 333–366. [CrossRef]
11. Rossi, V.; Cleeve Edwards, N.; Lundquist, L.; Schenker, U.; Dubois, C.; Humbert, S.; Jolliet, O. Life cycle assessment of end-of-life options for two biodegradable packaging materials: Sound application of the European waste hierarchy. *J. Clean. Prod.* **2015**, *86*, 132–145. [CrossRef]
12. Shamsuyeva, M.; Endres, H.-J. Plastics in the context of the circular economy and sustainable plastics recycling: Comprehensive review on research development, standardization and market. *Compos. Part C* **2021**, *6*, 100168. [CrossRef]
13. Swetha, T.A.; Bora, A.; Mohanrasu, K.; Balaji, P.; Raja, R.; Ponnuchamy, K.; Muthusamy, G.; Arun, A. A comprehensive review on polylactic acid (PLA)—Synthesis, processing and application in food packaging. *Int. J. Biol. Macromol.* **2023**, *234*, 123715. [CrossRef]
14. Giammaria, V.; Capretti, M.; Del Bianco, G.; Boria, S.; Santulli, C. Application of poly(lactic acid) composites in the automotive sector: A critical review. *Polymers* **2024**, *16*, 3059. [CrossRef] [PubMed]
15. Khouri, N.G.; Bahú, J.O.; Blanco-Llamero, C.; Severino, P.; Concha, V.O.C.; Souto, E.B. Polylactic acid (PLA): Properties, synthesis, and biomedical applications—A review of the literature. *J. Mol. Struct.* **2024**, *1309*, 138243. [CrossRef]
16. DeStefano, V.; Khan, S.; Tabada, A. Applications of PLA in modern medicine. *Eng. Regen.* **2020**, *1*, 76–87. [CrossRef] [PubMed]
17. Al-Shalawi, F.D.; Hanim, M.A.; Ariffin, M.K.A.; Kim, C.L.S.; Brabazon, D.; Calin, R.; Al-Osaimi, M.O. Biodegradable synthetic polymer in orthopaedic application: A review. *Mater. Today Proc.* **2023**, *74*, 540–546. [CrossRef]
18. Kharmanda, G. Challenges and Future Perspectives for Additively Manufactured Polylactic Acid Using Fused Filament Fabrication in Dentistry. *J. Funct. Biomater.* **2023**, *14*, 334. [CrossRef]
19. Charasseangpaisarn, T.; Wiwatwarrapan, C.; Thunyakitpisal, P.; Srimaneepong, V. Development of poly(methyl methacrylate)/poly(lactic acid) blend as sustainable biomaterial for dental applications. *Sci. Rep.* **2023**, *13*, 16904. [CrossRef]

20. Tsui, J.K.S.; Bell, S.; da Cruz, L.; Dick, A.D.; Sagoo, M.S. Applications of three-dimensional printing in ophthalmology. *Surv. Ophthalmol.* **2022**, *67*, 1287–1310. [CrossRef]
21. Zimina, A.; Senatov, F.; Choudhary, R.; Kolesnikov, E.; Anisimova, N.; Kiselevskiy, M.; Orlova, P.; Strukova, N.; Generalova, M.; Manskikh, V.; et al. Biocompatibility and physico-chemical properties of highly porous PLA/HA scaffolds for bone reconstruction. *Polymers* **2020**, *12*, 2938. [CrossRef]
22. Stodolak-Zych, E.; Ficek, K.; Wieczorek, J.; Kajor, M.; Gryń, K.; Rapacz-Kmita, A.; Rajca, J.; Kosenyuk, Y.; Stolarz, M.; Błażewicz, S. Assessment of sheep knee joint after ACL replacement with Achilles tendon autograft and PLA-based implant. *J. Mech. Behav. Biomed. Mater.* **2022**, *125*, 104923. [CrossRef]
23. Guerra, A.J.; Cano, P.; Rabionet, M.; Puig, T.; Ciurana, J. 3D-Printed PCL/PLA Composite Stents: Towards a New Solution to Cardiovascular Problems. *Materials* **2018**, *11*, 1679. [CrossRef]
24. Alhulaybi, Z.A. Fabrication and characterization of poly(lactic acid)-based biopolymer for surgical sutures. *ChemEngineering* **2023**, *7*, 98. [CrossRef]
25. Paz-González, J.A.; Velasco-Santos, C.; Villarreal-Gómez, L.J.; Alcudia-Zacarias, E.; Olivas-Sarabia, A.; Cota-Leal, M.A.; Flores-López, L.Z.; Gochi-Ponce, Y. Structural composite based on 3D printing polylactic acid/carbon fiber laminates (PLA/CFRC) as an alternative material for femoral stem prosthesis. *J. Mech. Behav. Biomed. Mater.* **2023**, *138*, 105632. [CrossRef] [PubMed]
26. Jin, Y.-A.; Plott, J.; Chen, R.; Wensman, J.; Shih, A. Additive Manufacturing of Custom Orthoses and Prostheses—A Review. *Proc. CIRP* **2015**, *36*, 199–204. [CrossRef]
27. Morimoto, S.Y.U.; Cabral, A.K.P.S.; Sanguinetti, D.C.M.; Freitas, E.S.R.; Merino, G.S.A.D.; Costa, J.Â.P.; Coelho, W.K.; Amaral, D.S. Upper limbs orthesis and prostheses printed in 3D: An integrative review. *Cad. Bras. Ter. Ocup.* **2021**, *29*, e2078. [CrossRef]
28. Tábi, T.; Ageyeva, T.; Kovács, J.G. Improving the ductility and heat deflection temperature of injection molded Poly(lactic acid) products: A comprehensive review. *Polym. Test.* **2021**, *101*, 107282. [CrossRef]
29. Freeland, B.; McCarthy, E.; Balakrishnan, R.; Fahy, S.; Boland, A.; Rochfort, K.D.; Dabros, M.; Marti, R.; Kelleher, S.M.; Gaughran, J. A review of polylactic acid as a replacement material for single-use laboratory components. *Materials* **2022**, *15*, 2989. [CrossRef]
30. RameshKumar, S.; Shaiju, P.; O'Connor, K.E.; Ramesh Babu, P. Bio-based and biodegradable polymers—State-of-the-art, challenges and emerging trends. *Curr. Opin. Green Sustain. Chem.* **2020**, *21*, 75–81. [CrossRef]
31. Mikula, K.; Skrzypczak, D.; Izydorczyk, G.; Warchoł, J.; Moustakas, K.; Chojnacka, K.; Witek-Krowiak, A. 3D printing filament as a second life of waste plastics—A review. *Environ. Sci. Pollut. Res.* **2021**, *28*, 12321–12333. [CrossRef]
32. DePalma, K.; Walluk, M.R.; Murtaugh, A.; Hilton, J.; McConky, S.; Hilton, B. Assessment of 3D printing using fused deposition modeling and selective laser sintering for a circular economy. *J. Clean. Prod.* **2020**, *264*, 121567. [CrossRef]
33. Rodríguez-Panes, A.; Claver, J.; Camacho, A.M. The influence of manufacturing parameters on the mechanical behaviour of PLA and ABS pieces manufactured by FDM: A comparative analysis. *Materials* **2018**, *11*, 1333. [CrossRef]
34. Alhazmi, M.W.; Backar, A.H. Influence of infill density and orientation on the mechanical response of PLA specimens produced using FDM 3D printing. *Int. J. Adv. Sci. Technol.* **2020**, *29*, 3362–3371.
35. Albadrani, M.A. Effects of raster angle on the elasticity of 3D-printed polylactic acid and polyethylene terephthalate glycol. *Designs* **2023**, *7*, 112. [CrossRef]
36. Zaneldin, E.; Ahmed, W.; Mansour, A.; Hassan, A.E. Dimensional stability of 3D printed objects made from plastic waste using FDM: Potential construction applications. *Buildings* **2021**, *11*, 516. [CrossRef]
37. *ISO 527:2019*; Plastics—Determination of Tensile Properties—Part 1: General Principles. ISO: Geneva, Switzerland, 2019.
38. Maszybrocka, J.; Dworak, M.; Nowakowska, G.; Osak, P.; Łosiewicz, B. The influence of the gradient infill of PLA samples produced with the FDM technique on their mechanical properties. *Materials* **2022**, *15*, 1304. [CrossRef]
39. Dave, H.K.; Rajpurohit, S.R.; Patadiya, N.H.; Dave, S.J.; Sharma, K.S.; Thambad, S.S.; Srinivasn, V.P.; Sheth, K.V. Compressive strength of PLA based scaffolds: Effect of layer height, infill density and print speed. *Int. J. Mod. Manuf. Technol.* **2019**, *11*, 21–27.
40. Hodzic, D.; Pandzic, A. Influence of infill design on compressive and flexural mechanical properties of FDM printed PLA material. *Ann. DAAAM Proc.* **2021**, *32*, 0191–0199.
41. Pandzic, A.; Hodzic, D. Mechanical properties comparison of PLA, tough PLA and PC 3D printed materials with infill structure—Influence of infill pattern on tensile mechanical properties. *IOP Conf. Ser. Mater. Sci. Eng.* **2021**, *1208*, 012019. [CrossRef]
42. Eryildiz, M. The effects of infill patterns on the mechanical properties of 3D printed PLA parts fabricated by FDM. *Ukr. J. Mech. Eng. Mater. Sci.* **2021**, *7*, 1–8. [CrossRef]
43. Torrente-Prato, G.; Sosa-Vivas, L.; Gonzalez-Delgado, J.; Hernandez-Silva, H.; León-Molina, H. Experimental and numerical study of the orthotropic behavior of 3D printed polylactic acid by material extrusion: Part two: An analysis about Poisson's ratios. *Prog. Addit. Manuf.* **2025**, *10*, 2335–2350. [CrossRef]
44. Wang, X.; Zhao, L.; Fuh, J.Y.H.; Lee, H.P. Effect of porosity on mechanical properties of 3D printed polymers: Experiments and micromechanical modeling based on X-ray computed tomography analysis. *Polymers* **2019**, *11*, 1154. [CrossRef]
45. Ferreira, R.T.L.; Amatte, I.C.; Dutra, T.A.; Bürger, D. Experimental characterization and micrography of 3D printed PLA and PLA reinforced with short carbon fibers. *Compos. Part B Eng.* **2017**, *124*, 88–100. [CrossRef]

46. Tüfekci, K.; Çakan, B.G.; Küçükakarsu, V.M. Stress relaxation of 3D printed PLA of various infill orientations under tensile and bending loadings. *J. Appl. Polym. Sci.* **2023**, *140*, e54463. [CrossRef]
47. Bertocco, A.; Bruno, M.; Armentani, E.; Esposito, L.; Perrella, M. Stress relaxation behavior of additively manufactured polylactic acid (PLA). *Materials* **2022**, *15*, 3509. [CrossRef] [PubMed]
48. Vanaei, S.; Rastak, M.; Vanaei, H.; Magri, E.; Tcharkhtchi, A.; Raissi, K. Orientation-dependent mechanical behavior of 3D printed polylactic acid parts: An experimental-numerical study. *Machines* **2023**, *10*, 2335–2350. [CrossRef]
49. Aly, R.; Olalere, O.; Ryder, A.; Alyammahi, M.; Samad, W.A. Mechanical property characterization of virgin and recycled PLA blends in single-screw filament extrusion for 3D printing. *Polymers* **2024**, *16*, 3569. [CrossRef]
50. Bergaliyeva, S.; Sales, D.L.; Delgado, F.J.; Bolegenova, S.; Molina, S.I. Manufacture and characterization of polylactic acid filaments recycled from real waste for 3D printing. *Polymers* **2023**, *15*, 2165. [CrossRef]
51. Gaweł, A.; Kuciel, S.; Liber-Kneć, A.; Mierzwiński, D. Examination of low-cyclic fatigue tests and Poisson's ratio depending on the different infill density of polylactide (PLA) produced by the fused deposition modeling method. *Polymers* **2023**, *15*, 1651. [CrossRef]
52. Gunasekaran, K.N.; Aravinth, V.; Kumaran, C.B.M.; Madhankumar, K.; Kumar, S.P. Investigation of mechanical properties of PLA printed materials under varying infill density. *Mater. Today Proc.* **2021**, *45*, 1849–1856. [CrossRef]
53. Ali, S.; Abdallah, S.; Devjani, D.; John, J.; Samad, W.; Pervaiz, S. Effect of build parameters and strain rate on mechanical properties of 3D printed PLA using DIC and desirability function analysis. *Rapid Prototyp. J.* **2022**, *29*, 92–111. [CrossRef]

materials

MDPI

Article

Multi-Objective Optimization of Injection Molding Process Parameters for Junction Boxes Based on BP Neural Network and NSGA-II Algorithm

Tengjiao Hong [1,2], Dong Huang [1], Fengjuan Ding [1,*], Liyong Zhang [1], Fulong Dong [1] and Lei Chen [3]

1 College of Intelligent Manufacturing, Anhui Science and Technology University, Chuzhou 233100, China; hongtengjiao@ahstu.edu.cn (T.H.); hdong12@126.com (D.H.); zhangly@ahstu.edu.cn (L.Z.); dongfl@ahstu.edu.cn (F.D.)

2 School of Business Administration, Stamford International University, Bangkok 10250, Thailand

3 Fengyang County Science and Technology Innovation Service Center, Chuzhou 233100, China; lchen2010@126.com

* Correspondence: dingfengjuan@ahstu.edu.cn; Tel.:+86-150-5610-9472

Abstract: Many factors affect the quality of the injection molding of plastic products, including the process parameters, mold materials, type and geometry of plastic parts, cooling system, pouring system, etc. A multi-objective optimization method for injection molding process parameters based on the BP neural network and NSGA-II algorithm is proposed to address the problem of product quality defects caused by unreasonable process parameter settings. Taking the junction box shell as the object, numerical simulation was carried out using Moldflow2019 software and a six-factor five-level orthogonal experiment was designed to explore the influence of injection molding process parameters, such as the mold temperature, melt temperature, injection pressure, holding pressure, holding time, and cooling time, on the volume shrinkage rate and warpage deformation of the junction box. Based on a numerical simulation, the BP neural network and NSGA-II algorithm were used to optimize the optimal combination of injection molding process parameters, volume shrinkage rate, and warpage deformation. The research results indicate that the melt temperature has the most significant impact on the quality of the injection molding of junction boxes, followed by the holding time, holding pressure, cooling time, injection pressure, and mold temperature. After optimization using the BP neural network and the NSGA-II algorithm, the optimal process parameter combination was obtained with a melt temperature of 230.03 °C, a mold temperature of 51.27 °C, an injection pressure of 49.13 MPa, a holding pressure of 69.01 MPa, a holding time of 15.48 s, and a cooling time of 34.91 s. At this time, the volume shrinkage rate and warpage deformation of the junction box were 6.905% and 0.991 mm, respectively, which decreased by 33.2% and 3.8% compared to the average volume shrinkage rate (10.34884%) and warpage deformation (1.030764 mm) before optimization. The optimization effect was significant. In addition, the errors between the volume shrinkage rate and warpage deformation predicted by BP-NSGA-II and the simulated values using Moldflow software were 1.9% and 3.4%, respectively, indicating that the optimization method based on the BP neural network model and NSGA-II algorithm is reliable.

Keywords: junction box; warping deformation; volume shrinkage rate; BP neural network; multi-objective optimization; process parameters

1. Introduction

Plastic has the advantages of stable performance and low density and is widely used in the information industry, aerospace, transportation, and other fields of the national economy [1]. It has become one of the important materials supporting the development of China's national economy and meeting people's daily life needs. Plastic Injection molding is widely used in the manufacturing of various plastic products due to its high production efficiency, low cost, and good product flexibility [2]. Plastic injection molding is a complex linear process with multiple parameters and interactions, which requires high requirements for process parameter settings [3,4]. Similar to traditional metal processing, plastic-injection-molded products also have defects, such as uneven shrinkage, weld lines, and warping deformation [5]. At present, mold manufacturers mostly obtain qualified plastic products by repeatedly repairing and modifying molds to address the problem of plastic product warping and deformation. This method of improving the quality of injection molded parts not only increases production costs but also takes a considerable amount of time, attention, and resources. With the surge in the demand for plastic products and rapid advances in the plastic industry, reducing the warping and deformation of plastic products during production has become an urgent need to be addressed in the injection mold industry.

Domestic and foreign scholars have conducted extensive research on plastic part warping deformation and have made significant progress. Li et al. [6] used Moldflow software and orthogonal experimental methods to study the effect of process parameters on the shrinkage rate of PFA (perfluoroalkoxy alkane) parts. The research results showed that the injection rate had the most significant impact on the shrinkage of the pipe length. In contrast, the melt temperature, holding pressure, and screw speed had the greatest impact on the shrinkage of the pipe's outer diameter. Guo et al. [7] used methods, such as Moldflow, DNN, RSM, and NSGA-II, to study the effects of factors, such as the mold temperature, solution temperature, packing pressure, packing time, injection molding time, cooling temperature, and cooling time, on the forming quality of polymer–metal hybrid (PMH) car front panels and obtained the optimal injection molding process parameters to guide actual production. The research results show that the proposed method has great potential in improving injection molding warpage deformation. Mukras et al. [8] aimed to minimize product defects (warpage deformation and volume shrinkage) and conducted actual injection molding experiments based on the face center composite design method. They analyzed the effects of seven process parameters, including the mold temperature, melt temperature, holding pressure, holding time, cooling time, injection speed, and injection pressure, on warpage deformation and volume shrinkage, and used a genetic algorithm to achieve multi-objective optimization. Zhou et al. [9] proposed a differential sensitivity fusion method (DSFM) that integrates a sampling strategy, numerical simulation, meta-modeling method, and a multi-objective optimization algorithm to achieve the multi-objective optimization of injection molding process parameters for automotive front bumpers. Ding et al. [10] proposed a multi-objective optimization method for the injection molding of thin-walled plastic parts based on tuna swarm optimization using a support vector machine (TSO-SVM) and multi-objective sparrow search algorithm (MOSSA), which solves the problems of warping deformation and volume shrinkage during the injection molding process. Nguyen et al. [11] combined orthogonal experiments, response surface methodology (RSM), and the NSGA-II method to optimize the quality process parameters of centrifugal pump casings during injection molding. Cao et al. [12] used simulation, the Latin hypercube sampling method, Bayesian optimized random forest regression (BO-RFR), gradient enhanced regression (GBR), and support vector regression (SVR) to construct a prediction model for process parameters, warpage, and volume shrinkage. Wang et al. [13] proposed a hybrid multi-objective optimization method that combines gradient-enhanced kriging (GEK) with multi-population differential evolution (MPDE) to minimize the warpage, volume shrinkage, and cycle time as optimization parameters. Wu et al. [14] proposed a multi-objective optimization method for the nonlinear shrinkage of micro-injection molded small modulus plastic gears based on a Moldflow simulation combined with a second-order response surface model and non-dominated sorting genetic algorithm II. The research results indicate that the most significant factor affecting the shrinkage of small modulus plastic gears is the holding time, and there are complex interactions between process parameters. Tan et al. [15] combined the Taguchi method with grey relational analysis to study the effects of the melt temperature, mold temperature, injection time, holding pressure/time, and cooling time on the size shrinkage and warping of wire harness connectors. Zeng et al. [16] proposed a multi-objective optimization method for injection molding process parameters based on hierarchical sampling and the comprehensive entropy weight to optimize the injection molding parameters of thin-walled propeller blades. The results showed that it was more optimized in multi-objective optimization than the response surface method. Hentati et al. [17] used the Taguchi method to study the effects of different molding parameters, such as the material temperature, injection pressure, holding time, and mold temperature, on the tensile stress (σ) and Young's modulus (E) of PC/ABS blends. The research results showed that the injection pressure and material temperature had the most significant impact on the mechanical properties and microstructure of PC/ABS blends. Guo et al. [18] aimed to optimize the production energy consumption, weight, and warpage. Based on Latin hypercube sampling, they established an optimal neural network model to predict the complex relationship between metal injection molding process parameters and the energy consumption, quality, and warpage. They used the NSGA-II algorithm and fuzzy decision-making based on criticism to achieve process parameter optimization. Liu et al. [19] took the warpage, clamping force, and sink marks as optimization objectives, and five process parameters, including the holding pressure, melt temperature, holding time, injection time, and melting time, were analyzed. They used the Taguchi orthogonal experiment, PSO-BP neural network model, improved particle swarm optimization algorithm, and TOPSIS method to achieve multi-objective optimization. Lee et al. [20] used an artificial neural network model to predict the nonlinear relationship between injection molding process parameters, such as the melt temperature, mold temperature, injection speed, holding pressure, holding time, cooling time, and product performance (quality, diameter, and height). Chang et al. [21] used the Pareto optimization framework and injection molding process parameters to perform multi-objective optimization based on the quality of drone shell parts. The research results showed that the injection time and pressure time are positively correlated with the mold indicators, with the strongest correlation, while the mold temperature shows a certain degree of a negative correlation. Li et al. [22] proposed an adaptive optimization method that combines the Design of Experiments (DOE) method with the Kriging surrogate model to reduce the warpage and residual stress of the bracket during injection molding. Yang et al. [23] used Moldflow software, uniformity testing, and the GA-BP neural network model to study the effects of five process parameters, including the mold temperature, melt temperature, holding pressure, injection time, and holding time, on the volume shrinkage rate and warpage variables of automotive plastic front frames. Gianluca et al. [24] adopted a central composite experimental design with flash formation and the component weight as optimization objectives, analyzing five process parameters, including the melting temperature, holding pressure, mold temperature, injection speed, and holding time. Regression equations and utility functions were used to achieve multi-objective optimization. Chen et al. [25] optimized the injection molding process parameters of plastic bottle preform parts based on the GA-BP-PSO algorithm. The research results showed that the warpage deformation obtained using the GA-BP-PSO algorithm differed little from the simulated values of Moldflow software. Li et al. [26] established a Kriging model based on orthogonal experiments and used numerical optimization algorithms, direct search methods, and global exploration methods to optimize the model, to obtain the optimal injection parameters with the minimum warpage deformation.

Based on the above literature, it can be concluded that using artificial neural network models, the NSGA-II algorithm, and the particle swarm optimization algorithm to study the optimization of plastic injection molding process parameters is reliable. To improve the quality and work efficiency of plastic injection molding, this study proposes a multi-objective optimization design method for the key process parameters of the plastic injection molding product quality, which combines the BP neural network and NSGA-II algorithm. The junction box shell is taken as the research object, and orthogonal experimental design and Moldflow simulation software are used for numerical simulation to obtain the original sample data. The parameters that have a significant impact on the quality of plastic parts are determined through variance analysis of the signal-to-noise ratio. Based on the BP neural network and NSGA-II, the injection molding process parameters of junction boxes are optimized to reduce the amount of warping deformation generated during the molding process and improve product quality.

2. Establishment of Finite Element Model and Process Analysis

2.1. Structural Analysis of Junction Box

The three-dimensional structure of the junction box housing is shown in Figure 1. As the chart shows, the shape of the junction box housing is irregular, with protrusions and holes. In the process of melt injection molding, it is necessary first to analyze the structure of the plastic parts and adopt reasonable injection methods to obtain higher-quality products. The plastic material is ABS, one of the widely used engineering plastics with excellent molding and mechanical properties. The recommended melt temperature is 200–280 °C, and the mold temperature is 25–80 °C [27].

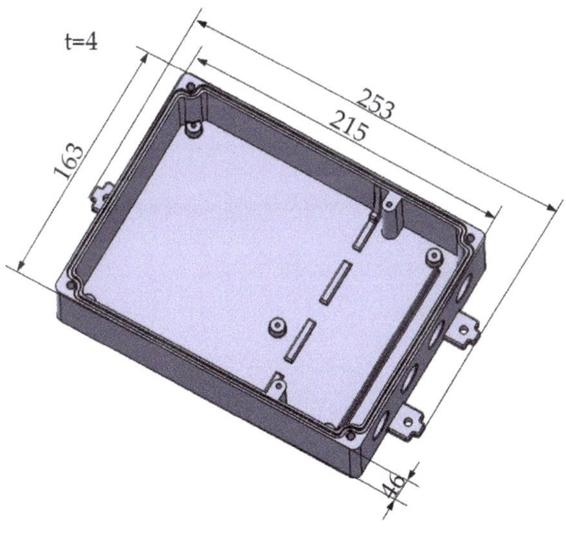

Figure 1. Three-dimensional structural diagram of junction box shell (size unit: mm).

2.2. Analysis of Mold Flow Under Initial Process Parameter Conditions

Grid division determines the accuracy of the discretization of finite element models and plays a crucial role in the accuracy and reliability of finite element analysis results. The neutral layer grid division speed is fast, but it cannot provide high accuracy; the Dual Domain with double-layer mesh is suitable for most product solutions and can provide a more accurate analysis and optimization; 3D solid meshes are often suitable for parts with complex structures and uneven thicknesses. A large number of meshes can lead to reduced efficiency and improved computational accuracy in finite element soft computing. Considering that the shape of the wiring housing is relatively simple and the wall thickness is relatively uniform in the overall structure, a double-layer mesh division method is adopted. Import the 3D model of the junction box housing into Moldflow software for automatic meshing. Pre-process the complex corners, sharp edges, and other structures of the injection molded parts through mesh repair to ensure that there were no multiple edges, free edges, intersecting elements, or completely overlapping elements. The mesh quality used for the simulation calculation is shown in Table 1, indicating that the mesh quality is good and there are no other defects. It is suitable for double-layer surface analysis and can be used for the Moldflow mold flow analysis simulation calculation.

Table 1. Grid quality of junction box shell.

Item	Actually	Requirement
Mesh-type	Triangles	—
Maximum aspect ratio	16.0	<20
Average aspect ratio	2.96	<3
Fully overlapping elements	0	0
Match percentage	86%	>85%
Total grids	51,388	—

The setting of the gate position of the junction box is related to the flow mode and flow time of the plastic melt in the mold [28]. Determine the optimal gate position for the junction box based on the Moldflow software, as depicted in Figure 2.

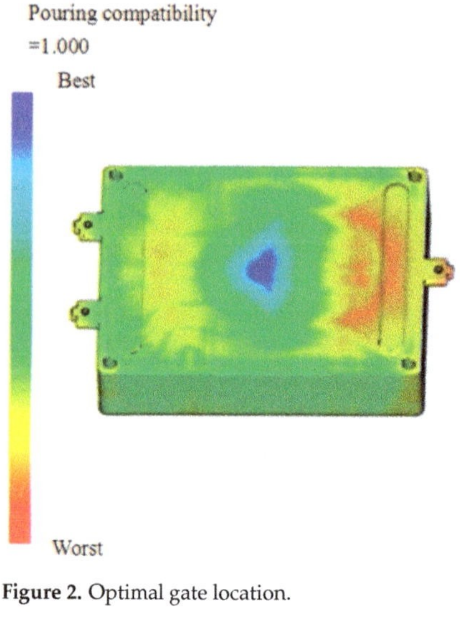

Figure 2. Optimal gate location.

According to the optimal gate position of the junction box shown in Figure 2, it can be seen that the blue area at the center of the plastic part has the best melt flowability and is more suitable as the optimal gate position. Considering the simple structure of the junction box plastic parts, to further reduce the number of defects, such as air pockets and weld lines that may be encountered during the analysis process, the final number of gates is determined to be 1.

The design of the pouring system has a significant impact on the forming quality of the junction box. Based on practical engineering experience, the first mock examination and two cavity pouring schemes are selected for the pouring system. To ensure the surface quality of the junction box shell, the gate position is fixed on the outside of the product. The cooling system can evenly cool the product after molding, reduce residual stress, maintain the dimensional accuracy of the plastic parts, and improve quality. The cooling system of the junction box adopts a simple and easy-to-process DC circulating cooling water channel, with specific parameters: the diameter of the water pipe is 10 mm, and the distance between the water pipe and the plastic parts of the parts is 20 mm. Establish a finite element model of the junction box for the pouring system and cooling system as shown in Figure 3.

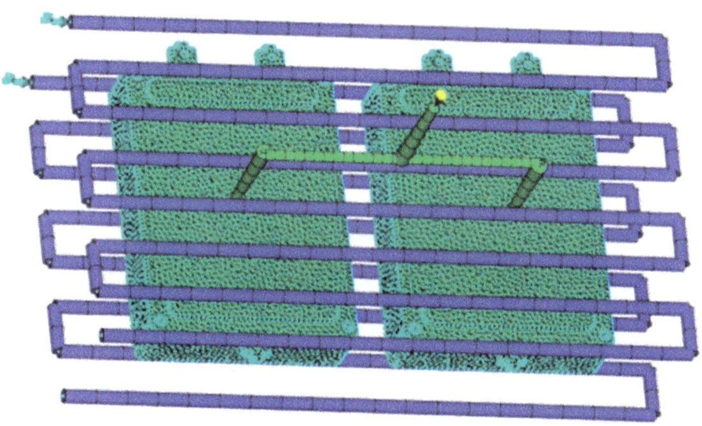

Figure 3. Overall finite element analysis model.

Determine the layout of the pouring and cooling system, conduct mold flow analysis on the junction box, set the analysis window to cooling + filling + holding pressure + warping, and set the process parameters to default values, as follows: the melt temperature is 240 °C, the mold temperature is 52.5 °C, the mold opening time is 5 s, the injection + holding pressure + cooling time is specified as 30 s, filling control is automatic, speed/pressure switching is automatic, the holding pressure control is filling pressure and time, without considering mold thermal expansion and angular effects. The analysis shows that the volume shrinkage rate and warpage deformation of the junction box are 10.35% and 1.352 mm, respectively.

3. Orthogonal Experiments

Considering that many factors affect the quality of the injection molding of junction boxes, it is difficult to measure them through process experiments or finite element simulation methods. To reduce computational costs and the number of experiments, the orthogonal experiments were used to optimize the experimental design. To investigate the influence of the injection molding process parameters on the warpage deformation of junction boxes, the melt temperature (A), mold temperature (B), injection pressure (C), holding pressure (D), holding time (E), and cooling time (F) were selected as factors for orthogonal experiments. Each factor was set at five levels to establish an orthogonal table of L25 (5^6). The experimental factors and factor levels are shown in Table 2.

Table 2. Factors and levels of orthogonal experiments.

Level \ Factor	Melt Temperature/°C	Mold Temperature/°C	Injection Pressure/MPa	Holding Pressure/MPa	Holding Time/s	Cooling Time/s
1	230	50	40	50	10	15
2	240	60	45	55	15	20
3	250	70	50	60	20	25
4	260	80	55	65	25	30
5	270	90	60	70	30	35

Using Moldflow simulation software to simulate the six-factor five-level orthogonal experiment described in Table 2, the evaluation indicators are the volume shrinkage rate (X_1/%) and maximum warpage (X_2/mm). The results of 25 sets of experiments are as described in Table 3.

Table 3. Results of orthogonal experiment.

Number	Experimental Factors						Experimental Result	
	A/°C	B/°C	C/MPa	D/MPa	E/s	F/s	X_1/%	X_2/mm
1	1	1	1	1	1	1	7.661	1.307
2	1	2	2	2	2	2	7.206	1.035
3	1	3	3	3	3	3	7.095	1.022
4	1	4	4	4	4	4	7.018	1.015
5	1	5	5	5	5	5	6.981	1.004
6	2	1	2	3	4	5	10.35	1.018
7	2	2	3	4	5	1	10.35	1.023
8	2	3	4	5	1	2	10.35	1.106
9	2	4	5	1	2	3	10.34	1.056
10	2	5	1	2	3	4	10.34	1.006
11	3	1	3	5	2	4	10.87	0.8403
12	3	2	4	1	3	5	10.87	1.008
13	3	3	5	2	4	1	10.87	1.015
14	3	4	1	3	5	2	10.88	1.016
15	3	5	2	4	1	3	10.88	1.281
16	4	1	4	2	5	3	11.41	1.013
17	4	2	5	3	1	4	11.42	1.012
18	4	3	1	4	2	5	11.41	0.9905
19	4	4	2	5	3	1	11.41	0.8618
20	4	5	3	1	4	2	11.41	0.9973
21	5	1	5	4	3	2	11.92	0.8007
22	5	2	1	5	4	3	11.92	0.8195
23	5	3	2	1	5	4	11.92	1.01
24	5	4	3	2	1	5	11.92	1.342
25	5	5	4	3	2	1	11.92	1.17

Based on the numerical simulation results of the injection molding of the junction box obtained from Table 3, the volume shrinkage rate and maximum warpage deformation degree of the junction box were statistically analyzed, and the average volume shrinkage rate and maximum warpage value of each factor and level were obtained. The signal-to-noise ratio range analysis was conducted on the volume shrinkage rate and warpage deformation of the junction box, and the results are shown in Tables 4 and 5. K_1–K_5 in Tables 4 and 5 represent the statistical mean values at different levels (1–5) and R represents the range of the statistical mean values. The larger the range, the more significant the impact of this factor on the examined indicators.

Table 4. Analysis of volume shrinkage ratio signal-to-noise ratio.

Mean Signal-to-Noise Ratio	Experimental Factors					
	A	B	C	D	E	F
K_1	−17.13	−20.28	−20.28	−20.27	−20.28	−20.28
K_2	−20.30	−20.17	−20.17	−20.17	−20.17	−20.17
K_3	−20.73	−20.14	−20.14	−20.15	−20.14	−20.14
K_4	−21.15	−20.12	−20.12	−20.13	−20.12	−20.12
K_5	−21.53	−20.11	−20.11	−20.11	−20.12	−20.11
R	4.39	0.16	0.16	0.16	0.16	0.16
Ranking	1	3	4	6	2	5

Table 5. Analysis of warpage ratio signal-to-noise ratio.

Mean Signal-to-Noise Ratio	Experimental Factors					
	A	B	C	D	E	F
K_1	−0.59547	0.16985	−0.14068	−0.58619	−1.60205	−0.54484
K_3	−0.19522	−0.23939	−0.28136	−0.38984	0.58246	−0.23929
K_4	0.23649	−0.40073	−0.50991	−0.09286	0.26778	0.22800
K_5	−0.06911	−0.71557	0.23794	0.72449	−0.11374	−0.54620
R	0.83196	0.92755	0.74785	1.35395	2.18451	0.77419
Ranking	4	3	6	2	1	5

In the study, the small size characteristic was used to analyze the difference between the volume shrinkage rate and warpage deformation. The larger the signal-to-noise ratio, the smaller the volume shrinkage rate and deformation obtained. From Tables 4 and 5, it can be seen that the factors affecting the volume shrinkage rate are A > E > B > C > F > D. The optimal process parameter combination is $A_1B_5C_5D_5E_5F_5$, while the factors affecting the maximum warpage amount are E > D > B > A > F > C, and the optimal process parameter combination is $A_4B_2C_5D_5E_3F_4$. By comparing with references [29–31], it can be seen that the influence of injection molding process parameters on the volume shrinkage rate and warpage deformation of plastic parts is different, and the process parameters also affect each other. Since ACB is the most significant factor affecting the volume shrinkage rate, $A_1C_5B_5$ is selected here; ED is the most significant factor affecting the maximum warpage deformation, and E_3D_5 is selected. The influence of F on the volume shrinkage rate is greater than the maximum warpage deformation, so F_5 is selected. Finally, the overall optimal combination is determined to be $A_1B_5C_5D_5E_3F_5$. By inputting the corresponding parameters into Moldflow software for simulation, a volume shrinkage rate of 7.128% and a maximum warpage deformation of 0.938 mm were obtained.

4. BP Neural Network

Due to the complex nonlinear relationship between injection molding process parameters and warpage deformation and the volume shrinkage of junction boxes, a BP neural network can be used to establish a prediction model for warpage deformation and the volume shrinkage of junction boxes under different process parameters. To accurately predict the warpage deformation and volume shrinkage of the junction box during the injection molding process, a three-layer BP neural network prediction model was constructed using six process parameters of injection molding as input variables and the volume shrinkage rate and warpage deformation as output variables. Therefore, the BP neural network has six input layer nodes and two output layer nodes, while the number of hidden layer nodes is mainly obtained through empirical formulas. The expression for a three-layer neural network is as follows [32]:

$$net_i = \sum_{j=1}^{n} x_i \omega_{ij} - \theta$$

$$y_i = f(net_i) \quad (1)$$

In Equation (1), x_i is the input layer variable; y_i is the output layer variable; ω_{ij} and θ represent the weights and thresholds of neurons, respectively.

In the experimental model, the S-type tangent function Tansig is used to verify the neuron activation function of the BP neural network, and the linear function Purelin is used as the neuron activation function in the output layer. The training function of the neural network is "Traingdm". To make the network converge quickly and solve more conveniently and avoid the phenomenon of neuron output saturation, the experimental data obtained through experiments will be normalized according to Equation (2) so that the data of the model training samples should be within the range of [−1, 1].

$$x = \frac{2 \times (x - x_{min})}{x_{max} - x_{min}} - 1 \quad (2)$$

where x is the raw data, y is the normalized result of x, x_{min} is the minimum value of the raw data, and x_{max} is the maximum value of the raw data.

Using 25 sets of orthogonal experimental results as sample data, randomly select the first 18 sets of samples for model training, and use the remaining sample data for model validation. The training objective of the BP neural network model is set to 10^{-5}, with a learning rate of 0.1, 100 iterations, and six validation failures. Using the coefficient of determination R^2 to determine the degree of fit of the regression equation in the training set, if R^2 is closer to 1, the linkage between the predicted and output results is closer, indicating a higher degree of fit of the neural network model. Using MATLAB 2019b software to train the BP neural network prediction model, it was found through trial and error that when the number of hidden layer nodes is five, the 6-5-2-2 three-layer BP neural network model established fits the best R^2 = 0.978. At this time, the BP neural network model predicts the volume shrinkage rate and maximum warpage deformation and compares it with the Moldflow simulation values as shown in Figure 4.

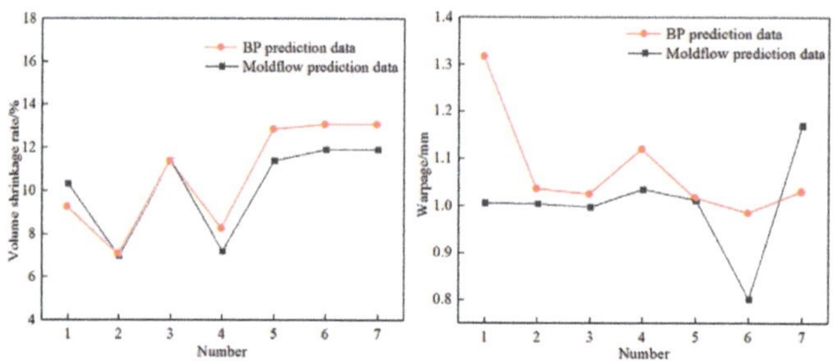

Figure 4. Comparison between BP neural network model predicted values and Moldflow simulation values.

According to Figure 4, it can be seen that the established BP neural network model is available to reflect the non-linear relation between the injection molding process parameters of the junction box and its volume shrinkage rate and warpage deformation. This neural network model has certain reliability and can be used for multi-target optimization of the injection molding process parameters of the junction box. Compared with references [19,20], it is feasible to establish an intrinsic mapping relationship between the molding quality and injection molding process parameters using the implicit statistical inference function of multi-layer feedforward neural networks, and the predicted results are relatively accurate.

The use of artificial neural network models to achieve intelligent process planning is of great significance for engineering applications.

5. Process Parameter Optimization Based on NSGA-II Algorithm

5.1. NSGA-II Algorithm

To obtain the optimal injection molding process parameters, the corresponding single objective quality prediction model is often established, and methods, such as the GA algorithm [8], PSO algorithm [33], etc., are commonly used to globally optimize the injection molding process. For the established multi-objective quality prediction model, the multi-objective particle swarm optimization algorithm [9], multi-objective genetic algorithm [34], NSGA-II algorithm [21], etc., are commonly used. The NSGA-II algorithm is an improvement based on the NSGA algorithm, which puts forward a fast and elite non-dominated sorting algorithm. It can not only reduce the computation but also merge the parent population with the child population, allowing the next generation to select from double the space and retain of all the most excellent individuals. The introduction of an elite retention strategy and competitive bidding mechanism ensures that certain excellent individuals in the population will not be discarded during the evolution process, thereby raising the precision of optimization results. By using a crowded-comparison operator, not only does it overcome the defect of manually specifying shared parameters in NSGA, but it also serves as a comparison standard between individuals in the population, allowing individuals in the quasi-Pareto domain to be uniformly extended to the entire Pareto domain, ensuring the diversity of the population. Using the NSGA-II algorithm to tackle the multi-objective problems is beneficial for obtaining the Pareto optimal solution with the minimum deviation between each objective function and the expected value.

5.2. Process Parameter Optimization

The nonlinear coupling model between the process parameters of the junction box constructed using the BP neural network and the volume shrinkage rate and warpage is used as the fitness function in the NSGA-II algorithm, and the technology parameters are civilized through the NSGA-II algorithm. The design variables and their ranges are shown below.

The melt temperature (A) is 230–270 °C, the mold temperature (B) is 50–90 °C, the injection pressure (C) is 40–60 MPa, the holding pressure (D) is 50–70 MPa, the holding time (E) is 10–30 s, and the cooling time (F) is 15–35 s. The optimization process of the BP neural network combined with the NSGA-II algorithm is shown in Figure 5.

When optimizing, the fitted BP neural network model is integrated into the main program of the algorithm in the form of an equation as the source of the initial population. The number of inputs is 6, the number of outputs is 2, the population size is set to 500, the crossover probability is set to 0.8, the mutation probability is set to 0.3, the maximum number of iterations is set to 50, and other parameters are set to default values. After 50 iterations of running the NSGA-II algorithm program in MATLAB, the Pareto solution was obtained as shown in Figure 6.

According to the optimization results shown in Figure 6, the algorithm has good convergence and can obtain a uniformly distributed Pareto optimal solution set. In this study, the multi-objective optimization model has two objective functions, and its Pareto front is a straight line. This indicates that the optimization trends of the volume shrinkage rate and warpage deformation of the junction box are opposite; that is, smaller warpage deformation corresponds to a larger volume shrinkage rate. It is difficult to obtain the optimal process parameters that achieve the minimum of both at the same time. In reference [19], the equal-weight TOPSIS method was used to obtain the optimal solution from the Pareto optimal frontier. In this study, the CRITIC comprehensive evaluation method was used to analyze and obtain the optimal process parameters, and both the volume shrinkage rate and warpage variables were found to be optimal. TOPSIS ranks each solution based on its proximity to the optimal and worst solutions, to determine the optimal design [35]. TOPSIS is suitable for multi-criteria decision-making and ranking problems. The CRITIC weighting method calculates and ranks the weights by constructing a judgment matrix and calculating consistency indicators. The CRITIC weighting method is suitable for multi-criteria evaluation and decision-making problems. The specific expression of the comprehensive evaluation model is as follows [36]:

$$Z = \sum_{j=1}^{2} W_j y_{ij} = W_1 y_{i1} + W_2 y_{i2} \tag{3}$$

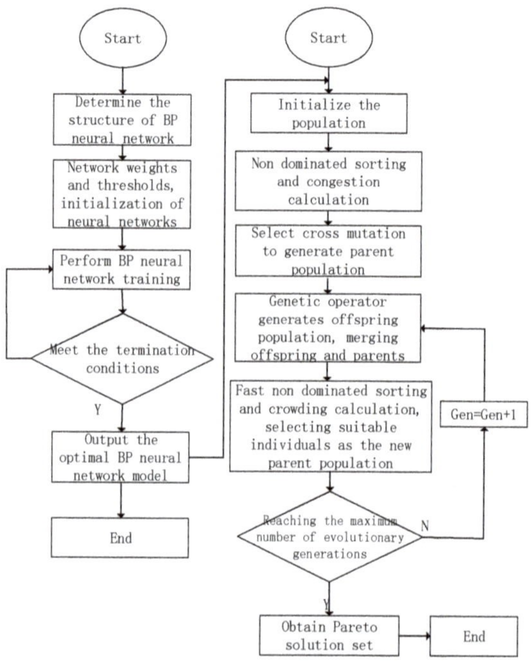

Figure 5. Optimization flowchart of BP neural network combined with NSGA-II algorithm.

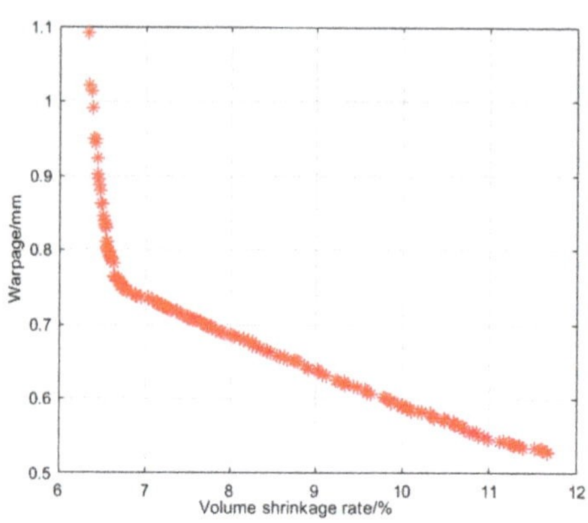

Figure 6. NSGA-II optimization results.

For the 50 sets of solutions obtained through NSGA-II multi-objective optimization, objective weights for the volume shrinkage rate and warpage deformation are calculated based on CRITIC. The general steps are as follows:

(1) Determine the indicator data matrix. There are a total of 50 samples to be evaluated, and the evaluation matrices for two evaluation indicators are as follows:

$$Y = y_{ij} = \begin{bmatrix} y_{11} & y_{12} \\ \vdots & \vdots \\ y_{50\times 1} & y_{50\times 2} \end{bmatrix} \quad (4)$$

Equation (4), y_{ij} represents the value of the jth evaluation index in the i-th experiment.

(2) Dimensionless processing.
To remove effects of different dimensions on the assessment, it is necessary to make the volume shrinkage rate and warpage deformation dimensionless. The smaller the indicators to be optimized, the better. Therefore, when standardizing, negative indicators are used for the calculation.

$$y'_{ij} = \frac{\left(y_j^{max} - y_{ij}\right)}{y_j^{max} - y_j^{min}} \quad (5)$$

Equation (5): y'_{ij} represents the standardized value, and y^{max} and y^{min} are the upper and lower limits of indicator j.

(3) Calculate indicator variability σ_j
In the CRITIC method, the standard deviation is used to represent the fluctuation of values within each indicator. The larger the standard deviation, the greater the numerical difference of the indicator and the stronger the evaluation strength of the indicator itself. Therefore, more weight should be assigned to the indicator.

$$\overline{y_j} = \frac{1}{n}\sum_{i=1}^{n} y'_{ij} \quad (6)$$

$$\sigma_j = \sqrt{\frac{\sum_{i=1}^{n}(y'_{ij} - \overline{y})^2}{n-1}} \quad (7)$$

In Equations (6) and (7), n is the total number of experimental groups, and σ_j represents the standard deviation of the jth indicator.

(4) Calculate indicator conflict R_j

$$R_j = \sum_{i=1}^{n}(1 - r_{ij}) \quad (8)$$

In Equation (8), r_{ij} represents the correlation coefficient between evaluation indicators i and j. The correlation coefficient is used to represent the correlation between indicators. The stronger the correlation with other indicators, the less conflicting the indicator is with other indicators, reflecting more similar information. The evaluation content that can be reflected is more repetitive, which weakens the evaluation strength of the indicator to a certain extent. Therefore, the weight assigned to the indicator should be reduced.

(5) Calculate the amount of information C_j

$$C_j = \sigma_j \times R_j \tag{9}$$

The larger the C_j, the greater the role of the jth evaluation indicator in the entire evaluation indicator system, and it should be distributed with more weight.

(6) Calculate objective weight W_j

$$W_j = \frac{C_j}{\sum_{j=1}^{n} C_j} \tag{10}$$

According to Equations (5)–(10) above, gradually calculate the objective weight values of the two indicators. As shown in Table 6, the objective weight of the volume shrinkage rate in the two molding masses is 0.4878, and the objective weight of warpage deformation in the two molding masses is 0.5122. Based on the obtained weight values, calculate the comprehensive evaluation value of the molding quality for each group of schemes according to Equation (1). Among them, the comprehensive evaluation value of the injection molding process scheme corresponding to a volume shrinkage rate of 6.905% and a warpage deformation of 0.991 mm reached 3.876, which is the minimum value among the 50 schemes. At this time, the optimal process parameter combination is the melt temperature of 230.03 °C, mold temperature of 51.27 °C, injection pressure of 49.13 MPa, holding pressure of 69.01 MPa, holding time of 15.48 s, and cooling time of 34.91 s.

Table 6. Calculation of objective weights for evaluation indicators.

Indicator	Variability	Conflict-Oriented	Information Content	Weight
volume shrinkage rate	0.255	1.960	0.500	48.78%
warpage	0.268	1.960	0.525	51.22%

Comparing the optimal results of the process parameters before optimization in Table 3, it can be seen that the NSGA-II algorithm improves the warpage deformation during the injection molding process of the junction box. The above process combination was applied to Moldflow for the numerical simulation calculation, and the calculated volume shrinkage rate was 7.037%, with a warpage deformation of 0.957 mm, as shown in Figure 7. The relative errors of the volume shrinkage rate and warpage deformation predicted by the NSGA-II algorithm were 1.9% and 3.4%, respectively, indicating a high degree of agreement between the two. This is similar to the findings of the literature [11,18,19,35]. By establishing a multi-objective optimization model and using the NSGA-II algorithm, the injection molding process parameters can be optimized to improve the molding quality of injection molded products.

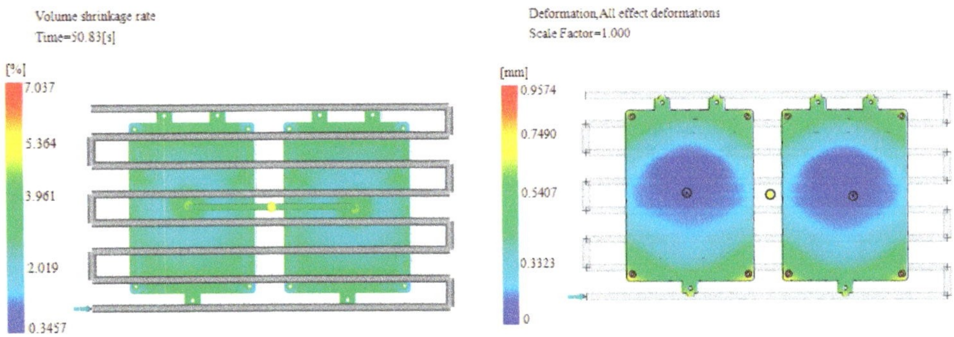

Figure 7. Numerical simulation results after algorithm optimization.

6. Conclusions

(1) Based on a six-factor five-level orthogonal experiment, the influence of the melt temperature, mold temperature, injection pressure, holding pressure, holding time, and cooling time on the volume shrinkage rate and warpage deformation of junction boxes was studied using Moldflow simulation software. By analyzing the range of the signal-to-noise ratio, the primary and secondary order of the influence of injection molding process parameters on the volume shrinkage rate and warpage deformation of junction boxes is obtained as follows: melt temperature > holding time > holding pressure > cooling time > injection pressure > mold temperature.

(2) Based on a fitted BP neural network model and combined with the NSGA-II algorithm, multi-target optimization was carried out within a specified parameter range using the quality indicators to determine the optimal solution, with a melt temperature of 230.03 °C, mold temperature of 51.27 °C, injection pressure of 49.13 MPa, holding pressure of 69.01 MPa, holding time of 15.48 s, and cooling time of 34.91 s. The resulting product had a warpage deformation of 0.912 mm, which was 33.2% lower than the average volume fraction before optimization (10.34884%). The optimal results were significant, and the injection molding quality of the junction box was improved.

(3) The errors between the volume shrinkage rate and warpage deformation predicted using BP-NSGA-II and the simulated values using Moldflow software are 1.9% and 3.4%, respectively, indicating that the optimization method of the BP neural network and NSGA-II algorithm is reliable.

Author Contributions: Writing—original draft preparation, conceptualization, T.H.; formal analysis and data curation, D.H.; software, writing—review and editing and funding acquisition, F.D. (Fengjuan Ding); visualization and project administration, L.Z.; visualization and methodology project administration, F.D (Fulong Dong); investigation and supervision, L.C. All authors have read and agreed to the published version of the manuscript.

Funding: This research was funded by the Talent Introduction Project of Anhui Science and Technology University (RCYJ202105); Design and Key Technology Research of Multi parameter Intelligent Control Instrument Junction Box (tzy202218); Natural Science Research Project of Higher Education Institutions in Anhui Province (2024AH050296); Research and Development of Fermentation Feed Drying Automatic Line (881314); Anhui Provincial Key Laboratory of Functional Agriculture and Functional Food, Anhui Science and Technology University (iFAST-2024-6); and Key technologies and applications of impinging stream based plant protection hedge spray system (2024AH050318).

Institutional Review Board Statement: Not applicable.

Informed Consent Statement: Not applicable.

Data Availability Statement: All data, models, and code generated or used during the study appear in the submitted article.

Acknowledgments: The authors are very grateful to Shubin Wang from the Anhui Jiuxing Instrument Company for providing us with the three-dimensional structural diagram of the junction boxes.

Conflicts of Interest: Author Lei Chen is employed by the company Fengyang County Science and Technology Innovation Service Center. The remaining authors declare that the research was conducted in the absence of any commercial or financial relationships that could be construed as a potential conflict of interest.

References

1. Wu, Y.; Feng, Y.; Peng, S.; Mao, Z.; Chen, B. Generative machine learning-based multi-objective process parameter optimization towards energy and quality of injection molding. *Environ. Sci. Pollut. Res.* **2023**, *30*, 51518–51530. [CrossRef] [PubMed]
2. Sun, X.; Zeng, D.; Tibbenham, P.; Su, X.; Kang, H.-T. A new characterizing method for warpage measurement of injection-molded thermoplastics. *Polym. Test.* **2019**, *76*, 320–325. [CrossRef]
3. Shen, C.; Wang, L.; Li, Q. Optimization of injection molding process parameters using combination of artificial neural network and genetic algorithm method. *J. Mech. Work. Technol.* **2007**, *183*, 412–418. [CrossRef]
4. Mohan, M.; Ansari, M.; Shanks, R.A. Review on the Effects of Process Parameters on Strength, Shrinkage, and Warpage of Injection Molding Plastic Component. *Polym. Technol. Eng.* **2017**, *56*, 1–12. [CrossRef]
5. Agrawal, A.R.; Pandelidis, I.O.; Pecht, M. Injection-molding process control—A review. *Polym. Eng. Sci.* **2010**, *27*, 1345–1357. [CrossRef]
6. Li, X.; Liu, F.; Zhang, C.; Gao, R.; Xu, X.; Fu, J.; Zhao, P. The influence of mold design and process parameters on dimensional shrinkage of perfluoroalkoxy alkane injection molding parts. *J. Appl. Polym. Sci.* **2024**, *141*, e55757. [CrossRef]
7. Guo, F.; Han, D.; Kim, N. Multi-Objectives Optimization of Plastic Injection Molding Process Parameters Based on Numerical DNN-GA-MCS Strategy. *Polymers* **2024**, *16*, 2247. [CrossRef]
8. Mukras, S.M.S.; Omar, H.M.; Al-Mufadi, F.A. Experimental-Based Multi-objective Optimization of Injection Molding Process Parameters. *Arab. J. Sci. Eng.* **2019**, *44*, 7653–7665. [CrossRef]
9. Zhou, H.; Zhang, S.; Wang, Z. Multi-objective optimization of process parameters in plastic injection molding using a differential sensitivity fusion method. *Int. J. Adv. Manuf. Technol.* **2021**, *114*, 423–449. [CrossRef]
10. Ding, W.; Fan, X.; Guo, Y.; Lu, X.; Wang, D.; Wang, C.; Zhang, X. Quality monitoring of injection molding based on TSO-SVM and MOSSA. *J. Polym. Eng.* **2024**, *44*, 64–72. [CrossRef]
11. Nguyen, H.T.; Nguyen, M.Q. A Numerical Simulation and Multi-objective Optimization for the Plastic Injection Molding of a Centrifugal Pump Casing. *IOP Conf. Ser. Earth Environ. Sci.* **2023**, *1278*, 012026. [CrossRef]
12. Cao, Y.; Fan, X.; Guo, Y.; Ding, W.; Liu, X.; Li, C. Multi-objective optimization of injection molding process parameters based on BO-RFR and NSGAII methods. *Int. Polym. Process.* **2023**, *38*, 8–18. [CrossRef]
13. Wang, Z.; Li, J.; Sun, Z.; Bo, C.; Gao, F. Multiobjective optimization of injection molding parameters based on the GEK-MPDE method. *J. Polym. Eng.* **2023**, *43*, 820–831. [CrossRef]
14. Wu, W.; He, X.; Li, B.; Shan, Z. An Effective Shrinkage Control Method for Tooth Profile Accuracy Improvement of Micro-Injection-Molded Small-Module Plastic Gears. *Polymers* **2022**, *14*, 3114. [CrossRef]
15. Tan, M.; Peng, S.; Huo, Y.; Li, M. Multi-Objective Optimization of Injection Molding Parameters for Manufacturing Thin-Walled Composite Connector Terminals. *Materials* **2024**, *17*, 3949. [CrossRef]
16. Zeng, W.; Yi, G.; Zhang, S.; Wang, Z. Multi-objective optimization method of injection molding process parameters based on hierarchical sampling and comprehensive entropy weights. *Int. J. Adv. Manuf. Technol.* **2024**, *133*, 1481–1499. [CrossRef]
17. Hentati, F.; Masmoudi, N. The impact of injection molding process parameters on mechanical properties and microstructure of PC/ABS blends using Taguchi approach. *Polym. Bull.* **2024**, *81*, 10659–10679. [CrossRef]
18. Guo, W.; Lu, T.; Zeng, F.; Zhou, X.; Li, W.; Yuan, H.; Meng, Z. Multi-objective optimization of microcellular injection molding process parameters to reduce energy consumption and improve product quality. *Int. J. Adv. Manuf. Technol.* **2024**, *134*, 5159–5173. [CrossRef]
19. Liu, F.; Pang, J.; Xu, Z. Multi-Objective Optimization of Injection Molding Process Parameters for Moderately Thick Plane Lens Based on PSO-BPNN, OMOPSO, and TOPSIS. *Processes* **2023**, *12*, 36. [CrossRef]
20. Lee, J.; Kim, J.; Kim, J. A Study on the Architecture of Artificial Neural Network Considering Injection-Molding Process Steps. *Polymers* **2023**, *15*, 4578. [CrossRef]
21. Chang, H.; Zhang, G.; Sun, Y.; Lu, S. Non-Dominant Genetic Algorithm for Multi-Objective Optimization Design of Unmanned Aerial Vehicle Shell Process. *Polymers* **2022**, *14*, 2896. [CrossRef] [PubMed]

22. Li, H.; Liu, K.; Zhao, D.; Wang, M.; Li, Q.; Hou, J. Multi-Objective Optimizations for Microinjection Molding Process Parameters of Biodegradable Polymer Stent. *Materials* **2018**, *11*, 2322. [CrossRef] [PubMed]
23. Yang, K.; Tang, L.; Wu, P. Research on Optimization of Injection Molding Process Parameters of Automobile Plastic Front-End Frame. *Adv. Mater. Sci. Eng.* **2022**, *2022*, 5955725. [CrossRef]
24. Trotta, G.; Cacace, S.; Semeraro, Q. Optimizing process parameters in micro injection moulding considering the part weight and probability of flash formation. *J. Manuf. Process.* **2022**, *79*, 250–258. [CrossRef]
25. Chen, Z.; Hu, H.; Shen, J.; Ni, D.; Wang, Z. Optimization of warpage deformation of thin-walled injection molded. *Eng. Plast. Appl.* **2024**, *52*, 70–75. (In Chinese)
26. Li, S.; Fan, X.Y.; Guo, Y.H.; Liu, X.; Huang, H.Y.; Cao, Y.L.; Li, L.L. Optimization of Injection Molding Process of Transparent Complex Multi-Cavity Parts Based on Kriging Model and Various Optimization Techniques. *Arab. J. Sci. Eng.* **2021**, *46*, 11835–11845. [CrossRef]
27. Fang, M.; Zhang, Y.; Wang, C.; Wang, R. Optimization of Injection Process Parameters for Automotive Instrument Framework Based on Particle Swarm Optimization Algorithm. *China Plast. Ind.* **2024**, *52*, 79–85+90. (In Chinese)
28. Ding, Y.; Hu, S. Numerical Simulation and Process Optimization of Bottle Cap Injection Molding Based on Orthogonal Experiment. *Plast. Sci. Technol.* **2018**, *46*, 91–94. (In Chinese)
29. Sun, Z.; Li, J.; Wang, Z.; Bo, C.; Zhang, Y.; Yao, K.; Gao, D. Multi-objective optimal design of key process parameters for plastic injection molding product quality. *J. Chem. Eng. Chin. Univ.* **2023**, *37*, 483–491. (In Chinese)
30. Rosli, M.; Ishak, M.I.; Jamalludin, M.R.; Khor, C.; Nawi, M.; Syafiq, A.M. Simulation-Based Optimization of Plastic Injection Molding Parameter for Aircraft Part Fabrication Using Response Surface Methodology (RSM). *IOP Conf. Ser. Mater. Sci. Eng.* **2019**, *551*, 012108. [CrossRef]
31. Chen, W.C.; Nguyen, M.H.; Chiu, W.H.; Chen, T.N.; Tai, P.H. Optimization of the plastic injection molding process using the Taguchi method, RSM, and hybrid GA-PSO. *Int. J. Adv. Manuf. Technol.* **2016**, *83*, 1873–1886. [CrossRef]
32. Ding, F.J.; Jia, X.D.; Hong, T.J.; Xu, Y.L. Flow Stress Prediction Model of 6061 Aluminum Alloy Sheet Based on GA-BP and PSO-BP Neural Networks. *Rare Met. Mater. Eng.* **2020**, *49*, 1840–1853.
33. Bensingh, R.J.; Machavaram, R.; Boopathy, S.R.; Jebaraj, C. Injection molding process optimization of a bi-aspheric lens using hybrid artificial neural networks (ANNs) and particle swarm optimization (PSO). *Measurement* **2019**, *134*, 359–374. [CrossRef]
34. Feng, Q.; Liu, L.; Zhou, X. Automated multi-objective optimization for thin-walled plastic products using Taguchi, ANOVA, and hybrid ANN-MOGA. *Int. J. Adv. Manuf. Technol.* **2020**, *106*, 559–575. [CrossRef]
35. Zhang, T.; Wei, Z.; Wang, L.; Xue, Z.; Wang, S.; Wang, P.; Qi, B.; Yue, Z. Multi-objective optimization of composite stiffened panels for mass and buckling load using PNN-NSGA-III algorithm and TOPSIS method. *Thin-Walled Struct.* **2025**, *209*, 112878. [CrossRef]
36. Diakoulaki, D.; Mavrotas, G.; Papayannakis, L. Determining objective weights in multiple criteria problems: The critic method. *Comput. Oper. Res.* **1995**, *22*, 763–770. [CrossRef]

materials

MDPI

Article

A New Double-Step Process of Shortening Fibers without Change in Molding Equipment Followed by Electron Beam to Strengthen Short Glass Fiber Reinforced Polyester BMC

Michael C. Faudree [1,*] and Yoshitake Nishi [2]

1 Faculty of Liberal Arts and Science, Tokyo City University, Yokohama-shi 224-8551, Japan
2 Graduate School of Engineering, Tokai University, Hiratsuka-shi 259-1292, Japan; west@tsc.u-tokai.ac.jp
* Correspondence: faudree@tcu.ac.jp

Abstract: It is vital to maximize the safety of outdoor constructions, airplanes, and space vehicles by protecting against the impact of airborne debris from increasing winds due to climate change, or from bird strikes or micrometeoroids. In a widely-used compression-molded short glass fiber polyester bulk-molded compound (SGFRP-BMC) with 55% wt. $CaCO_3$ filler, the center of the mother panel has lower impact strength than the outer sections with solidification texture angles and short glass fiber (SGF) orientations being random from 0 to 90 degrees. Therefore, a new double-step process of: (1) reducing commercial fiber length without change in molding equipment; followed by a (2) 0.86 MGy dose of homogeneous low-voltage electron beam irradiation (HLEBI) to both sides of the finished samples requiring no chemicals or additives, which is shown to increase the Charpy impact value (a_{uc}) about 50% from 6.26 to 9.59 kJm^{-2} at median-accumulative probability of fracture, $P_f = 0.500$. Shortening the SGFs results in higher fiber spacing density, S_f, as the thermal compressive stress site proliferation by action of the CTE difference between the matrix and SGF while the composite cools and shrinks. To boost impact strength further, HLEBI provides additional nano-compressive stresses by generating dangling bonds (DBs) creating repulsive forces while increasing SGF/matrix adhesion. Increased internal cracking apparently occurs, raising the a_{uc}.

Keywords: SGFRP; BMC; impact strength; fiber length; texture; electron beam irradiation

1. Introduction

With the increase in frequency and intensity of disaster events from climate change, it is important to advance materials that can withstand the increasingly harsh environmental conditions that can occur. It is crucial to always strengthen materials for maximum safety, with the utmost concern for the environment. Bulk molded compounds (BMCs) have been widely-used across many industries for light-load bearing articles, such as electrical housing, car headlights, and appliance parts. BMCs have advantages over metals in being corrosion resistant and lightweight when used as non-structural parts for aircraft or other vehicles to lower fuel consumption and reduce CO_2 emissions. Other advantages are their easy formability of complex shaped parts, and their resistance to hot or cold for house appliances and outdoor household articles. BMCs usually have fiber content of ~5 to 30 wt.% [1–12] which includes glass [11,12], carbon [13], as well as jute [14] and kenaf [15]. Short glass fiber reinforced polymer–bulk molded compounds (SGFRP-BMCs) are 3-phase fiber-filler-polymer systems constructed with fiber at ~5 to 30 wt.% [1,3,7] and filler of $CaCO_3$ at ~35 to 55 wt.% [7,8]. For BMCs in general, fillers of TiO_2, Al_2O_3, SiC, $Mg(OH)_2$, ZnO [9], fumed silica [10], fly ash [5], or waste thermosetting BMC [2] have been used.

Up to now, the combination of shortening fibers and electron beam treatment to enhance mechanical properties of fiber reinforced polymers (FRPs) has not been found in the literature. However, other strengthening methods have always been advancing for

FRPs [16–25]. For example, for long fiber FRP, pre-stressing fibers before intrusion with the polymer melt has been a useful tool to increase mechanical properties [16–19]. Similar to pre-stressing steel bars in concrete, Pang and Fancey found viscoelastic prestressing of multi-filament nylon 6,6 yarn in a bisphenol-A-based low viscosity epoxy resin increased tensile strength, modulus, and toughness up to 15%, 30%, and 40%, respectively [16]. For a unidirectional glass fiber reinforced polymer (GFRP) epoxy composite, Hadi and Ashton found pre-stressing GFs at 25, 50, 75, 100, and 200 MPa increased tensile strength and elastic modulus [17]. Motahari and Cameron found prestressing fibers increased flexural properties of FRP [18]. Interestingly, Jenkins et al. found by controlling prestressing magnitude and eccentricity, mold-free FRP composites' internal stress conditions can be manipulated to obtain curved part geometry with high precision [19]. Another method commonly used for strengthening FRPs has been enhancing adhesion at the fiber/matrix interface [20–25]. Yuan et al. found that applying good a coupling agent to GFs increased fracture stress, but decreased fracture strain in polyvinyl chloride (PVC) GFRP [20]. Meraghni et al. modelled the effect of interfacial degradation on a short fiber-reinforced polymer containing matrix microcracks [21]. Numerous fiber treatments have been applied to inert CF to increase its weak adhesion to polymer matrix [22–25], including electrochemical modification [22], electro-polymer coating [23], plasma surface modification [24], and Ni sputtering [25].

Past research has shown the mechanical properties of 2-phase fiber/polymer FRP systems that shortening fiber length decreases mechanical properties such as impact strength, tensile stress, and strain [26–34], and that longer fibers are desired [35]. For example, in a polypropylene (PP) GFRP at GF wt.% from 3 to 60%, stiffness was found to be lower at shorter fiber lengths below 0.5 mm, and virtually unchanged above 0.5 mm [26]. Above 40 wt.% GF content, modulus was lowered by fiber packing problems and increased in voids [26]. Another study of PP GFRP showed that impact strength was increased as GF length was raised to 6.4 mm, with a strain energy model predicting optimal length to be 8 mm [28]. In a PP carbon fiber reinforced polymer (CFRP), a sequential reduction in mechanical properties, Izod impact, tensile strength and modulus, bending strength and modulus, and Rockwell hardness were found as CF length was reduced from 10 to 5 to 2 to 1 to 0.5 mm [32]. In a CFRP, tensile strength and stiffness were increased by lengthening CF from 2 to 4 mm, but were decreased as CF length was increased further to 6.4 mm. Optimal fiber length was reported to be 4 mm [33]. Numerical modeling has also been carried out on 2-phase GFRP systems showing that as GFs are lengthened, strength increases rapidly at low fiber lengths, especially near the critical length, l_c of ~1.0 mm, and flattens out at about $5l_c$ [29,30]. Fu and Lauke calculated l_c to be 0.56–0.59 mm for nylon FRP, 1.4 mm for polypropylene (PP) FRP, and 0.68 to 0.84 mm for polybutylene terephthalate (PBI) FRP [29]. For 2-phase systems, below l_c, fibers do not impart stress transfer to the matrix, i.e., they are too short to exhibit shear lag with the matrix during tension. Hence, instead of breakage, the dominant fracture mechanism for fibers below l_c is pull-out, weakening the 2-phase fiber/polymer system. However, with the 3-phase fiber-filler-polymer system, CTE difference is the dominant mechanism, where shortening fibers to 0.44 mm strengthens, rather than weakens, the composite [36]. For green composites also (2-phase), a trend was found, being that lowering fiber length decreases mechanical strength, such as those with hemp fiber (HF) [34,37], jute fiber (JF) [38,39], sisal fiber (SF) [40], or agave fiber (AF) [41]. For example, in HF-reinforced thermoplastic polyurethane (HFRP), tensile strength was raised from 16 to peak out at 27 MPa by lengthening HFs from 6 mm to 15 mm [34]. However, above 15 mm to 40 mm, tensile strength remained approximately constant [34]. In injection-molded poly[styrene-b-(ethylene-co-butylene)-b-styrene) (SEBS)/HFRP composite with 30 wt.% HF), increasing nominal fiber length from 1.10 mm to 4.19 mm (from 0.57 mm to 1.03 mm after injection molding) significantly raised tensile strength from ~33 to ~39 MPa, and tensile modulus from ~2.3 to ~2.9 GPa [37]. Sajin et al. found for a compression molded alkali-treated JF-reinforced isopthalic polyester composite (JFRP), 20 mm fiber length gave the highest tensile strength and modulus, flexural strength and modulus, and impact strength. The 20 mm fiber length samples had higher mechanical properties than either the

5, 10, 15, or 25 mm samples [39]. In a compression-molded sisal fiber-reinforced PP (SFRP) composite with 40 wt.% SF, increasing chopped SF length from ~5 mm to ~25 mm (from ~3 mm to ~17 mm after compounding) resulted in increase in tensile strength from ~36 to ~43 MPa, flexural strength from ~41 to ~62 MPa, flexural modulus from 1.4 to ~2.9 GPa, and impact strength from 3.25 to 4.09 kJm^{-2} [40]. Moreover, for a biodegradable composite fabricated from agave leaves and epoxy (AFRP), longer AF lengths of 60 mm were found to give better tensile and flexural properties than those that were 10 mm or chopped [41]. Hence, in 2-phase fiber-polymer systems, mechanical properties typically decrease with a decrease in fiber length.

However, counter to the results stated in the literature for numerous 2-phase fiber/polymer composite systems, in the 3-phase SGFRP-BMC, several mechanical properties have been increased by decreasing fiber length below that of commercially used samples [7,8,36]. For a widely-used compression-molded styrene butadiene SGFRP-BMC with 11 wt.% SGF and 55 wt.% $CaCO_3$ filler, it was found that shortening SGF length from nominal (commercial) 6.4 mm to submillimeter 0.44 mm raised impact strength 25% at the typically weak center of the mother panel [36]. Moreover, for an injection-molded styrene-butadiene SGFRP-BMC with 20 wt.% SGF, and 47.1 wt.% $CaCO_3$ filler, shortening SGFs from 6.4 mm to 0.44 mm increased tensile modulus 5 to 25% [8]. Ultimate tensile strength (UTS) and its strain were also boosted by ~60 and ~40%, respectively [7]. This was unusual, since there is typically a tradeoff: as UTS is enhanced, its strain is typically decreased [16]. Other mechanical tests, such as flexural and fatigue, would provide a more thorough characterization, but are beyond the scope of this study.

In the above cases, SGFs were shortened by 30 min in an extended mix of the paste in a sigma-blade mixer prior to molding without a change in molding equipment. Enhancements by shortening fibers has been described with a "fiber spacing" model [8] depicted in Figure 1. For the simple case when fibers are separated by spaces in a given length, x, number of fibers, N_f will be equivalent to number of spaces, S_f, as shown in Equation (1):

$$S_f = N_f \tag{1}$$

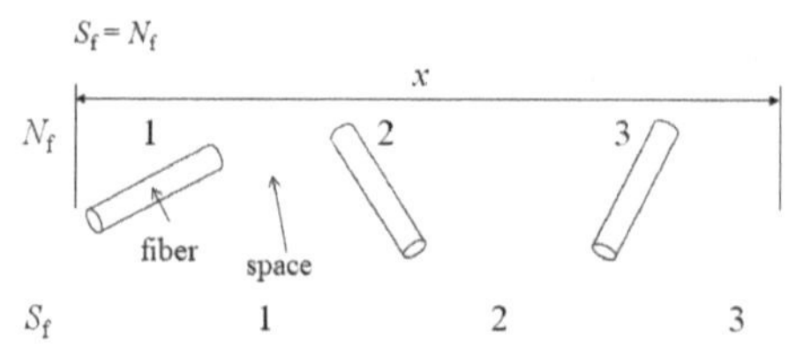

Figure 1. Fiber spacing model adapted from Faudree et al. (2014) [8].

This would be independent of fiber orientation, θ, with respect to x. And apply to the random fiber orientation of the SGFRP-BMC panel center. Moreover, since both N_f and S_f are dimensionless quantities, they can be put into any dimensional coordinate system; hence, S_f is given in three dimensions as "fiber spacing density" (mm^{-3}).

Enhancements of mechanical properties were attributed to the increase in S_f as mean fiber length, l_f (mm) is shortened, acting to increase the micro-compressive stress sites of the matrix on the SGFs by a difference in coefficient of thermal expansion (CTE) during cooling down and shrinking [8]. S_f is related to l_f by the following Equation [8]:

$$S_f = \rho_f = V_f/(\pi r^2 l_f) \tag{2}$$

where V_f and r are SGF volume fraction and mean fiber radius (0.007 mm). It follows that the CTE of cured polyester resin matrix is 55 to 100 $\times$ 10^{-6}/K [42], an order of magnitude above E-glass fibers at 5.4 $\times$ 10^{-6}/K [43]. In concert with this, since the SGFRP-BMC is a three-phase fiber-filler-polymer system, the filled matrix behaves as a polymer-filler subsystem in the narrow spacing between fibers that efficiently allows for an increase in mechanical properties by decreasing the fiber length. During cooling down and shrinking, the polymer compresses onto the hard $CaCO_3$ filler with a high surface area of the 1 to 7 μm $CaCO_3$ particles, whose listed CTE is ~4.6×10^{-6}/K [44]. As S_f is increased by reducing SGF length, micro-compressive stress sites are increased, raising the mechanical properties.

Note that in the SGFRP-BMC under study, V_f of hard components of SGFs plus $CaCO_3$ filler is quite high at 46%, with the polymer mixture at 54%. Small particles that are highly dispersed and in close proximity are excellent for maximizing thermal residual stresses [45] for strengthening the BMC composites. Another point here is in the case of boosting the ultimate tensile stress and strain; the increase in S_f also acts to halt cracks before propagating above the critical length for the SGFRP-BMC to take on more of the load [7]. Therefore, for the 3-phase SGFRP-BMC, shortening the fiber length has been found to increase mechanical properties.

Other studies on shorter fiber lengths increasing mechanical properties are nearly non-existent. However, Senthilrajan et al. found, for a jute reinforced polyester (JFRP), at JF content of 25 wt.%, 5 mm fiber length samples had higher flexural strength and modulus, and a specific strength and modulus than those with longer lengths of 10, 15, 20, or 25 mm [38]. Their scanning electron microscope (SEM) analysis attributed the strengthening to stronger JF/Matrix bonding [38].

As mentioned previously, the $CaCO_3$ filler particles of the SGFRP-BMC appear to act to raise tensile strength and its strain, along with tensile modulus, and impact strength when SGF length is reduced [7,8]. Metal matrix composites (MMC) and cemented carbide composites, as well as ceramics, are reported to exhibit this strengthening mechanism, in that reduced particle size increases mechanical properties [46–48]. Improvements were attributed to increased residual compressive stress sites by the CTE mismatch between the particles and the matrix. In the SGFRP-BMC, $CaCO_3$ filler particles were measured using SEM to be in about the same size range, <1 to 7 μm [8], as that for ceramics at <1 μm to several microns [46–48].

Concerning the influence of filler on BMCs, little research has been performed in the literature. An important study for sustainability for hard-to-recycle BMCs was conducted by Matykiewicz et al., who crushed waste thermosetting BMC, using it as reinforcement in an epoxy composite [2]. Lautenschlägera et al. investigated the effects of filler for 25 vol.% jute chopped fiber-reinforced BMCs and needle-punched non-woven SMCs [14]. They showed that $CaCO_3$ filler was found to produce higher tensile strength and Young's modulus than that of paltry kaolin. In contrast, they found flexural strength had high scatter for either filler, due to inhomogeneous fiber distribution of fiber-rich and -sparse volumes [14]. Investigations on the effect of percent filler is beyond the scope of this study.

In injection-molded SGFRP-BMC, the mean SGF length of 30 min extended mix samples was measured using SEM to be sub-millimeter at 0.44 mm (std. dev. = ±0.203 mm) [1,7]. Thus, it is assumed, for the compression molded samples, that the 30 min mixing reduces SGF length to about 0.44 mm. With standard deviation of ±0.203 mm, SGF length difference between the 30 min extended mix compression-molded samples and commercial 6.4 mm samples is considered enough to obtain reliable results [7,8]. Notably, the sub-millimeter length is below that of the estimated l_c of ~1.0 mm reported for 2-phase FRPs [29,30], and l_c would depend on adhesion strength at the fiber/matrix interface. The extended mixing of 30 min is reported to have insignificant effect on the $CaCO_3$ filler particles (<1 to ~7 μm) [2].

To boost impact strength further, homogeneous low-voltage electron beam (HLEBI) treatment to finished specimens with the shortened 0.44 mm SGFs will be employed. This is because, previously, 0.86 MGy HLEBI treatment to commercial 6.4 mm samples was found to increase impact values from 5 to 25% [49]. SEM showed that 0.86 MGy HLEBI resulted in much more polymer/$CaCO_3$ matrix adhering to the SGFs than that of untreated, which were left virtually bare. An electron spin resonance (ESR) analysis of the samples showed HLEBI generated a strong peak (inflection point at 323.0 mT) signifying the presence of dangling bonds (DBs). The DBs are lone pair electrons in the outer orbital shells that exhibit repulsive forces acting to increase internal micro-compressive stress sites, and strengthen the SGF/Matrix interface [49]. This allows for energy dispersion in the form of higher numbers of cracks generated in the SGFRP-BMC samples to take on higher energy impacts [49]. An examination of tested samples showed that the fracture mechanism transitioned at a_{uc} of 5.4 to 6.7 kJm^{-2} from clean to smaller secondary micro-crack proliferation, sometimes including bends near the main crack. Fracture surface area was observed to increase with increasing impact strength. The conversion was independent of HLEBI-treated or untreated conditions [49].

HLEBI has a positive track record of enhancing GFRPs with long fibers, GFs themselves, silica glass, and several types of polymers [49,50]. HLEBI is also reported to strengthen the fiber/matrix interface with electronic charge generation and the formation of DBs [49].

HLEBI typically severs bonds with the lowest BDE (Figure 2a,b) for polymeric components in the SGFRP-BMC. DBs are formed at the terminated atoms with low BDE [49] in methylene groups (CH_2–CH_2: 369 kJmol^{-1}), hydroxyl groups (CH_3–OH: 377 kJmol^{-1}), and phenyl groups (C_6H_5–R: <377~389 kJmol^{-1} where R– is CH_3– or C_2H_5–). Free-radical hydrogens have higher BDE (H–CH: 427 kJmol^{-1}, H–OC_2H_5: 435 kJmol^{-1}, and H–$OOCCH_3$: 469 kJmol^{-1}). Alkene groups H_2C=CH_2 have the highest BDE at 720 kJmol^{-1} [51,52].

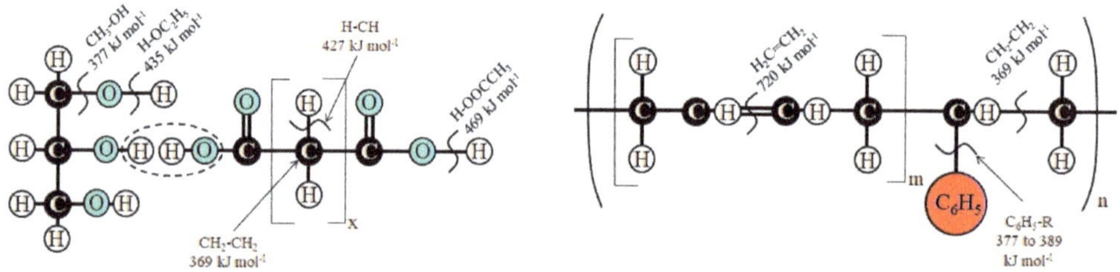

(**a**) thermoset polyester (**b**) styrene-butadiene

Figure 2. Depiction of HLEBI-activated: (**a**) thermoset polyester, and (**b**) styrene-butadiene copolymer, respectively, adapted from Faudree et al. (2012) [49], indicating approximate bond dissociation energies (BDE) [51,52] and DBs.

The action of HLEBI results in atoms being spaced further apart by DB generation, creating compressive stress sites by repulsion of the generated lone-pair electrons. For silica glass, for instance, in the radial distribution function (RDF), the normalized coordination number (N_D/N_O) is reduced, while the normalized mean atomic distance, (r_D/r_O) is increased [49]. This results in nano-scale volume expansion that can act to increase mechanical properties.

As mentioned earlier, we strengthen the weak center of the SGFRP-BMC panel by shortening SGFs followed by HLEBI. This is because separate studies of shortening SGFs from commercial 6.4 mm to 0.44 mm [36]; and applying 0.86 MGy HLEBI to finished 6.4 mm samples [49] were found to successfully raise impact energy of the weak center of the SGFRP-BMC mother panel, 25%, and 5 to 25%, respectively. Hence, it seems possible that the combination can additively boost impact strength by 50%. Therefore, we chose the independent parameters of (1) 6.4 mm SGFs, (2) 0.44 mm, and (3) 0.44 mm + 0.86 HLEBI to illustrate the enhancements. The 6.4 mm is chosen since it is widely used commercially, while the 0.44 mm was found to yield significant increase in mechanical properties [7,36]. In addition, the 0.86 MGy HLEBI dose was chosen since preliminary testing showed that it produced the highest impact values out of a wide range of dose levels [49]. Therefore, the goal of this study is to apply a double-step process of (1) shortening the nominal 6.4 mm fiber length formulation to 0.44 mm by 30 min extended mixing without change in molding equipment, followed by (2) applying a 0.86 MGy dose of HLEBI to both sides of the SFGRP-BMC finished samples with no chemicals or additives to boost impact values at the weak center of the mother panel.

2. Materials and Methods

2.1. Preparation of SGFRP-BMC Samples

Compression-molded SGFRP-BMC panels were provided by Premix, Inc., (now Citadel Plastics) North Kingsville, OH, USA, with components listed in Table 1. A strong coupling agent was used on the SGFs, but its components are proprietary. Processing parameters, panel geometry, and volume fractions are listed in Table 2. A schematic of the compression molding process is illustrated in Figure 3, where SGFs are mixed into a paste prior to injection molding. The test matrix is composed of three data sets listed in Table 3 designated as "6.4 mm", "0.44 mm", and "0.44 mm + HLEBI" samples. The "6.4 mm" data set is the commercial formulation SGFRP-BMC with nominal 6.4 mm SGFs which underwent 20 min mixing of the paste with a double-arm sigma blade Banbury mixer before injection molding. The "0.44 mm" are those with an extended mix of 30 min, equaling 50 min mixing total. Mean SGF length was found to be 0.44 mm [36]. The 6.4 mm and 0.44 mm samples were not treated with HLEBI. A third data set, the "0.44 mm + HLEBI", was treated with 0.86 MGy HLEBI on both sides of the finished samples.

Table 1. Weight percents of SGFRP-BMC components.

COMPONENT	Wt.%
Propylene glycol maleate polyester	13.75
Styrene butadiene copolymer	12.75
Commercial E-glass fibers	11
$CaCO_3$ filler	55
Aluminum silicate filler	3
Magnesium hydroxide	0.5
Proprietary initiators and inhibitors	4

Table 2. Processing parameters.

PARAMETER	Condition
Mold pressure	5.5 to 6.9 MPa (800 to 1000 psi)
Temperature	422 K (149 °C)
Cure time	2 min
Mold type	Matched metal die compression mold
Panel size	304.8 × 304.8 mm (12 × 12 in)
Panel thickness	2 mm
V_f (SGF)	0.080
V_f ($CaCO_3$)	0.377
V_f (remaining polymer mixture)	0.543

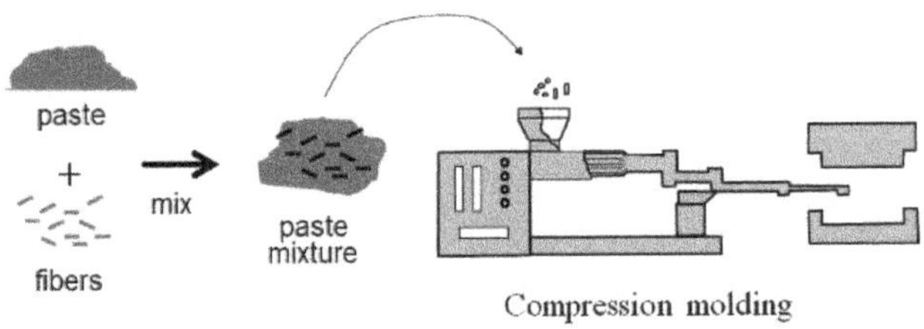

Figure 3. Schematic of compression molding process.

Table 3. Data sets of the SGFRP-BMC.

Experimental Condition	6.4 mm Untreated	0.44 mm Untreated	0.44 mm + HLEBI
Number of Specimens	14	14	14

Figure 4 shows the fabrication steps for the three data sets, (a) 6.4 mm, (b) 0.44 mm, and (c) 0.44 mm + HLEBI. The commercial 6.4 mm and 0.44 mm samples were fabricated in three and four steps, respectively [36]. Figure 4 shows that the 0.44 mm + HLEBI samples were fabricated in five steps. Step 1 is mixing the paste for 20 min. Step 2 is 30 min of extended mixing. Step 3 is compression molding the paste into panels. Step 4 is cutting the samples. Finally, Step 5 is applying 0.86 MGy HLEBI to both sides of the samples (see the HLEBI Section).

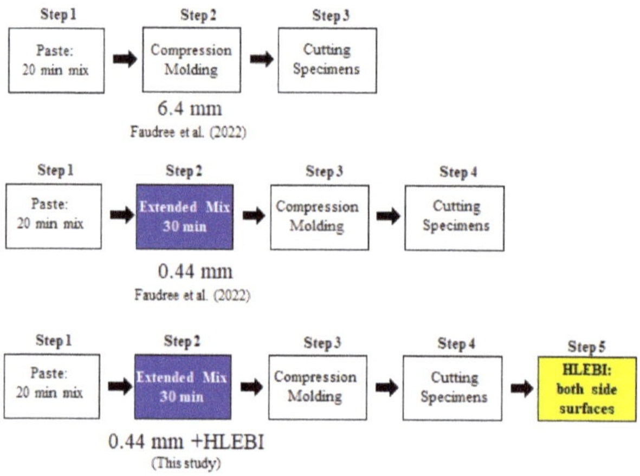

Figure 4. Steps of SGFRP-BMC fabrication for.4 mm; 0.44 mm (Faudree et al. (2022)) [36]; and 0.44 mm + HLEBI samples, respectively.

Figure 5 is an illustration of the SGFRP-BMC panels, showing specimens taken from the weak center section. During compression molding of the paste charge, the direction of flow is outward, but more random just below the plunger in the center. After molding, specimens (80 × 10 × 2 mm) were cut according to ASTM D 6110-02 (2002) [53] for anisotropic panels [53]. Since this study focusses on improving impact strength at the weak panel center, samples cut from outer 3 sub-quadrants of each quadrant are not shown. Only the 4 sub-quadrants at center are shown, and are assumed to have identical flow patterns since they exhibited statistically lower impact strength than the outer sections [1,36,49]. Locations of samples cut from the panel center are indicated in Figure 5. For consistency, specimens in each sub-quadrant are numbered from 1 to 7 outward from the center. The 14-sample data sets were taken from two sub-quadrants, two each from sample numbers 1 to 7.

2.2. Homogeneous Low-Voltage Electron Beam Irradiation

After molding and cutting, samples were treated with a HLEBI processor (Table 4) (Type LB250/15/180L, Energy Science, Inc., Woburn, MA, USA, Iwasaki Electric Group, Ltd., Tokyo, Japan), as shown in Figure 6a,b [49]. Faudree et al. (2022) gives a detailed explanation [36]. To estimate how far into sample thickness HLEBI activates, penetration depth, D_{th} (μm) is calculated. When ρ is sample density (g cm^{-3}) and V is acceleration voltage at sample surface (kV), the D_{th} can be obtained [54]:

$$D_{th} = 66.7V^{5/3}/\rho \quad (3)$$

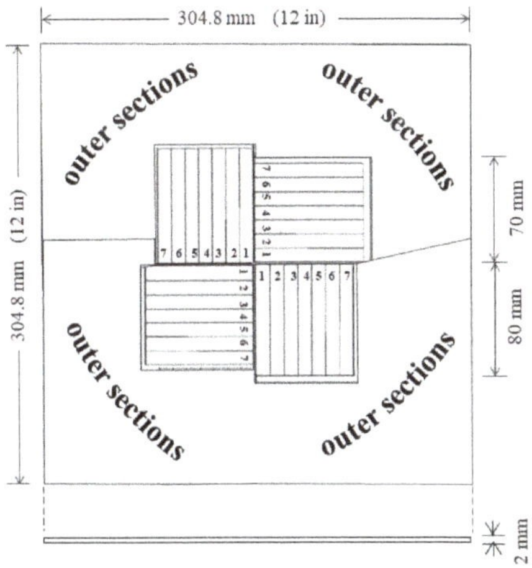

Figure 5. Schematic of Charpy impact sample cutting from the weak center section of a SGFRP-BMC panel. Specimen numbers indicated. (outer section specimens not shown).

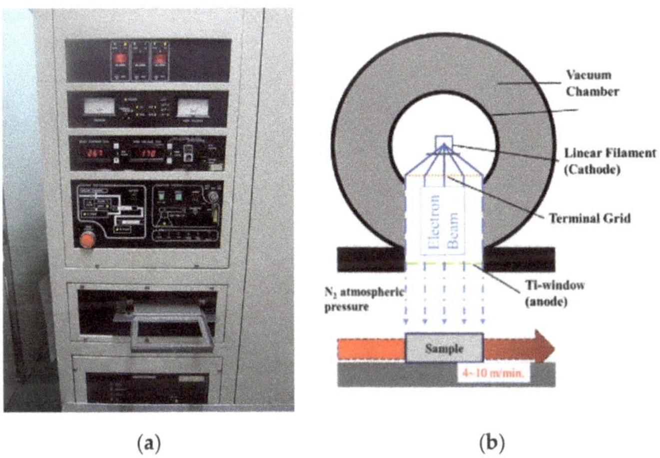

Figure 6. Electron-curtain processor (Iwasaki): (**a**) photo, and (**b**) schematic.

Table 4. HLEBI parameters.

Acceleration voltage	170 kV
Irradiating current	2.68 mA
Irradiation environment	N2 gas atmosphere
Residual O_2 conc.	<300 ppm
Conveyor speed	10 m min^{-1}
EB dose/sweep	0.0432 MGy
Sweep time (one way)	0.20 s

Table 4. *Cont.*

Gap interval bet. sweeps	20 s
EB yield calc.	FWT Nylon dosimeter

Individual D_{th} are listed in Table 5. D_{th} for the SGFRP-BMC is 116 μm, or about 5.8% into the thickness. Since both specimen sides are HLEBI-activated, there is a skin/core/skin sandwich structure of 0.116/1.87/0.116 mm. Impact tests were carried out 30 ± 0.5 h after HLEBI irradiations.

Table 5. Densities and the calculated D_{th} for the SGFRP-BMC (bold type) and its individual components [54].

Component	**Density, ρ (g cm^{-3})**	**D_{th} (μm)**
Polymer Mixture	1.200	185
Matrix Compos.	1.847	120
SGFRP-BMC	**1.917**	**116**
SGF	2.620	84.5
$CaCO_3$ Filler	2.800	79.1

2.3. Charpy Impact Tests

Charpy impact strength was measured using a drop-weight pendulum apparatus (Shimadzu Corp. No. 51735, Tokyo, Japan) [55,56] according to the JIS K 7077-1991 testing standard [55]. Figure 7a,b shows the impact tester and specimen mount. Specimen dimensions were 80 × 10 × 2 mm. Testing parameters and details can be found in Faudree et al. (2022) [36].

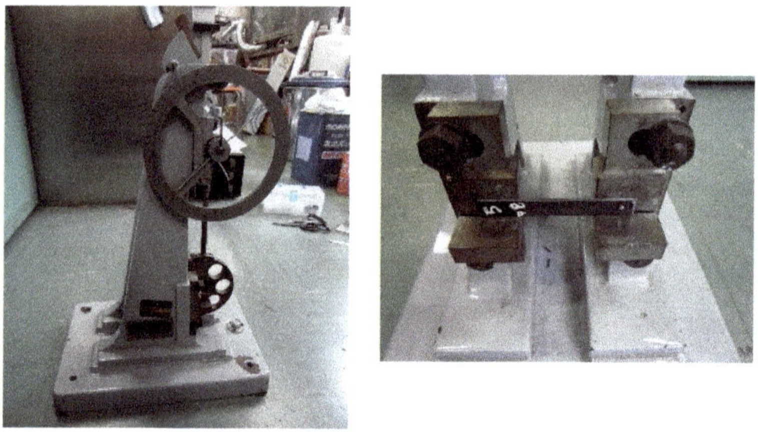

(**a**) (**b**)

Figure 7. Photos of: (**a**) Charpy impact tester, and (**b**) mount with SGFRP-BMC specimen.

3. Results

Effect of Shortening SGFs and HLEBI on Impact Values

Experimental results are shown in Figure 8a,b for the 6.4 mm (black dots), 0.44 mm (blue triangles) [36], and 0.44 mm + HLEBI (yellow squares) samples, respectively. Figure 8a shows accumulative probabilities (P_f) vs. Charpy impact value (a_{uc}) according to a general form of the median rank method [57] described in detail in Faudree et al. (2022) [36]. Figure 8a shows, for the 0.44 + HLEBI data set, our double-step process of (1) shortening fibers to 0.44 mm with 30 min extended mixing, followed by (2) HLEBI of 0.86 MGy, boosted a_{uc} at all P_f over that of the 6.4 mm. The 0.44 mm data set (without HLEBI, blue triangles) is from a previous study reported by Faudree et al. (2022) [36], and is shown here to illustrate that the 0.44 mm + HLEBI data set yields higher a_{uc} than the 0.44 mm at all P_f. Although at low-P_f of 0.049, a_{uc} of the 0.44 mm data set was reduced (4.06 kJm^{-2}) from that of the 6.4 mm (4,.92 kJm^{-2}) [36], the 0.44 mm + HLEBI process showed a significant improvement to 7.06 kJm^{-2}. As stated in [36], the a_{uc} are calculated from ($a_{uc} = E/(bt)$) with impact energy, and E is obtained with ($E = WR[(\cos\beta - \cos\alpha) - (\cos\alpha' - \cos\alpha)(\alpha + \beta)/(\alpha - \alpha')]$). The P_f are calculated from ($P_f = (I - 0.3)/(N_s + 0.4)$).

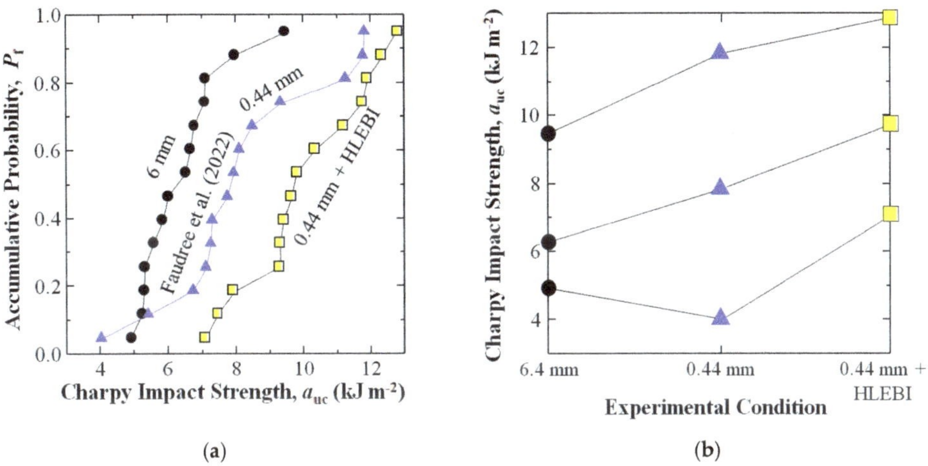

Figure 8. Experimental results showing: (**a**) changes in Charpy impact values a_{uc} (kJm^{-2}) for SGFRP-BMC samples of untreated 6 mm and 0.44 mm, along with 0.44mm + HLEBI (0.86 MGy), with (**b**) the a_{uc} from (**a**) a_{uc} at low-, median-, and high-P_f of 0.049, 0.500, and 0.951, respectively. The 0.44 mm data are from Faudree et al. (2022) [36].

Figure 8b shows a_{uc} at low-, median-, and high-P_f of 0.049, 0.500, and 0.951, respectively. Namely, the 0.44 mm + HLEBI data set shows significant increase in a_{uc} over that of 6.4 mm. At P_f of 0.049, 0.500, and 0.951, a_{uc} was increased 43%, 55%, and 35%, respectively, from 4.92 to 7.06 kJm^{-2}, 6.26 to 9.72 kJm^{-2}, and 9.45 to 12.80 kJm^{-2}. For the 0.44 mm data set, Figure 8b shows the increases in a_{uc} at median- and high-P_f, but a decrease at low-Pf of 0.049 compared with those of 6.4 mm [36]. However, the 0.44 mm + HLEBI process significantly raises the a_{uc} over those of the 6.4 mm samples at low-P_f of 0.049, along with those at median-, and high P_f.

Table 6 provides individual a_{uc} for each sample of the data sets in Figure 8.

Table 6. Charpy impact values, a_{uc} (kJm^{-2}), of individual specimens along with their accumulative probabilities, P_f.

P_f	6.4 mm Untreated	0.44 mm Untreated	0.44 mm+ HLEBI
0.049	4.92	4.06	7.06
0.118	5.24	5.44	7.45
0.188	5.29	6.74	7.91
0.257	5.32	7.12	9.28
0.326	5.57	7.26	9.30
0.396	5.84	7.31	9.41

Table 6. *Cont.*

P_f	6.4 mm Untreated	0.44 mm Untreated	0.44 mm+ HLEBI
0.465	6.01	7.76	9.64
0.535	6.51	7.96	9.80
0.604	6.64	8.12	10.33
0.674	6.76	8.51	11.16
0.743	7.06	9.35	11.75
0.813	7.09	11.26	11.89
0.882	7.97	11.79	12.33
0.951	9.45	11.84	12.80

4. Discussion

4.1. Average Impact Strength

Since average values are commonly employed in strength evaluations, Figure 9 is included here showing the average a_{uc} ($a_{uc,avg}$) with standard deviation bars. Figure 9 shows the double-step process of 0.44 mm + HLEBI significantly increased $a_{uc,avg}$ 56% over that of 6.4 mm from 6.41 to 10.00 kJm^{-2}. Although standard deviations were quite high at 1.24 and 1.81 kJm^{-2} for the 6.4 and 0.44 + HLEBI data sets, respectively, the lower limit of the 0.44 mm + HLEBI (8.19 kJm^{-2}) was higher than the upper limit of the 6.4 mm (7.65 kJm^{-2}). This provides further support that our double-step process can apparently increase the a_{uc} of weak center sections of the SGFRP-BMC panels.

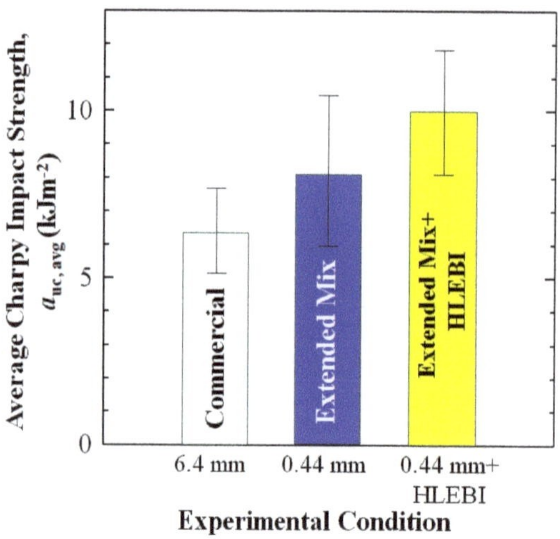

Figure 9. Average impact strength, $a_{uc,\,avg}$ (kJm^{-2}) with standard deviations (bars) for the three data sets: 6.4 mm, 0.44 mm, and 0.44 mm + HLEBI, respectively.

4.2. Micro-Strengthening Mechanism Using Extended Mix

Figure 10a–c illustrates the micro-strengthening mechanism by an extended mix of the SGFRP-BMC for the 6.4 mm, 0.44 mm, and 0.4 mm + HLEBI specimens. The SGFs are depicted as having a random orientation as in the panel center. Impact strength enhancements are attributed to increase in S_f and N_f, as l_f is shortened, acting to increase micro-compressive stress sites of the matrix on the SGFs by a difference in CTE during cooling down and shrinking [8,36]. The composite is hardened, increasing impact resistance.

Figure 11b shows an order of magnitude increase in S_f and N_f over that of 6.4 mm (a) that is exceptionally more dispersed. Here, N_f is made to approximate the actual situation where a total length of 58 of 0.44 mm SGFs equals 4 of 6.4 mm SGFs. To discuss enhancement during impact, Figure 10 is made to depict tensile side of specimen with impact area across specimen width (dotted lines). Shortening SGFs to 0.44 mm allows for a higher proportion of SGFs to cross the line, along with increased compressive stress sites (arrows), to counter tensile deformation from impact for higher a_{uc}. On the other hand, in Figure 10a, the less dispersed 6.4 mm samples have gaps in the form of areas lacking SGFs with less residual stresses resulting in easier crack initiation and propagation in the matrix at the tensile side when impacted.

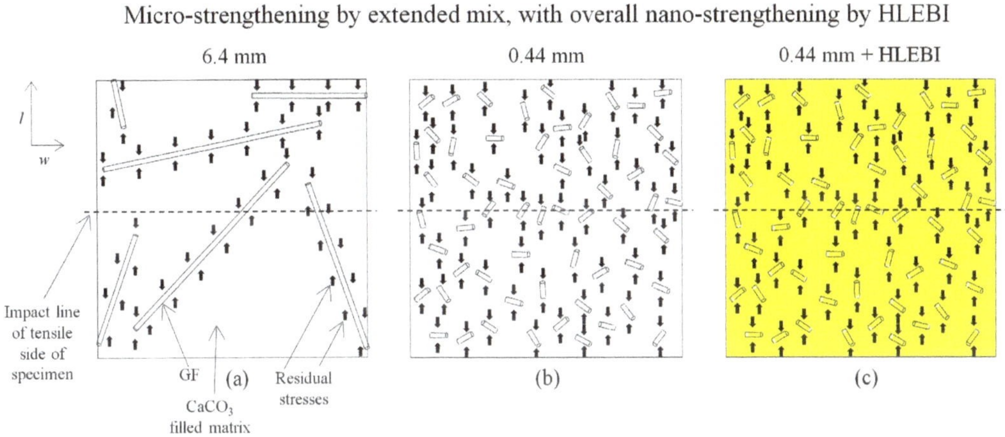

Figure 10. Schematic of micro-scale strengthening mechanism showing (**a**) 6.4 mm; (**b**) 0.44 mm; and (**c**) 0.44 mm + HLEBI samples, respectively. In (**c**), yellow indicates HLEBI activation. Specimen tensile sides are depicted with line of impact. Specimen length and width directions are indicated.

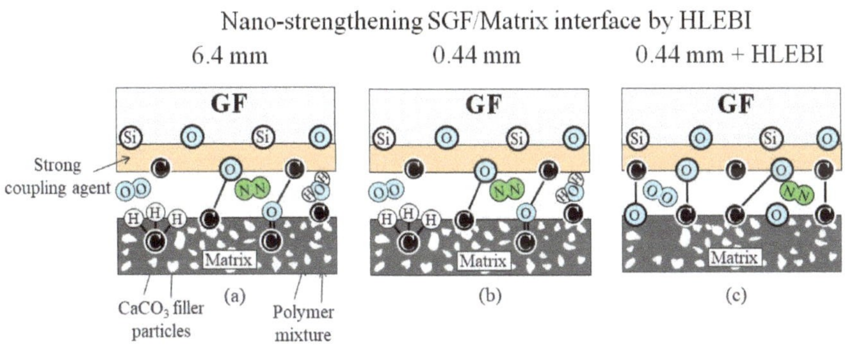

Figure 11. Schematic of the nano-scale SGF/Matrix interface-strengthening mechanism by HLEBI depicting an increase in strong bonds for: (**a**) 6.4 mm; (**b**) 0.44 mm; and (**c**) 0.44 mm + HLEBI samples, respectively.

As mentioned earlier, the filled matrix behaves as a polymer-filler subsystem in the spacing between fibers allowing for an increase, but not a decrease, in mechanical properties with decreasing fiber length. To increase a_{uc} further, Figure 10c depicts the nano-scale strengthening of the 0.44 mm composite with HLEBI represented in yellow.

4.3. *Nano-Strengthening Mechanism Using HLEBI*

To illustrate the nano-strengthening mechanism, action of HLEBI occurs at: (1) the SGF/Matrix interface; (2) within the SGFs; and (3) within the polymer matrix. These are explained in Figures 11 and 12, and previously in Figure 2a,b. HLEBI works by severing bonds, creating DBs in the form of lone pair electrons that enhance interfacial adhesion and strengthen bulk materials. When 0.86 MGy HLEBI dose is applied to both surfaces of finished SFGRP-BMC samples at the weak panel center, a_{uc} can be raised. Figure 11a,b shows, for untreated 6.4 mm and 0.44 mm, that the SGF/Matrix interface has typical strong chemical bonds between the SGFs and the matrix from the coupling agent. They are apparently accompanied by weak Van der Waals forces with trace atmospheric gasses, O_2, N_2, and H_2O existing at the interface. However, Figure 11c depicts that when HLEBI is applied in the 0.44 mm + HLEBI samples, additional strong bonds of C-C and C-O are apparently formed by lone electron pairs generated, i.e., DBs, raising the a_{uc}. As mentioned earlier, DBs have been detected in 0.86 MGy HLEBI-treated 6.4 mm fiber length SGFRP-BMC as peak generation using ESR analysis in a previous study [49]. Moreover, SEM revealed the HLEBI increased matrix adhering to SGFs, with increased impact values ~5 to 25% [49].

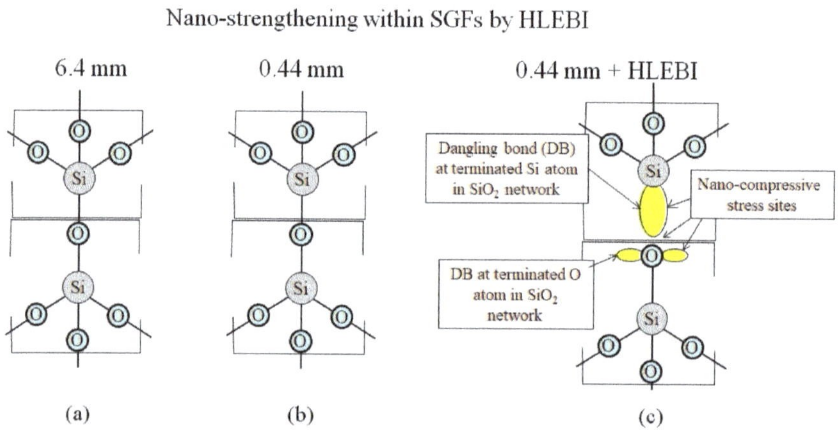

Figure 12. Schematic of the nano-scale strengthening mechanism within SGFs by HLEBI for: (**a**) 6.4 mm; (**b**) 0.44 mm; and (**c**) 0.44 mm + HLEBI samples, respectively.

Figure 12a–c illustrates the strengthening within SGFs by HLEBI. HLEBI has been reported to strengthen silica glass [49]. Figure 12a,b shows the unactivated bonds in the SGFs of 6.4 mm and 0.44 mm samples. However, Figure 12c shows HLEBI activation creates DBs in the outer shell electrons at terminated O atoms in the SiO_2 network in the 0.44 mm + HLEBI samples. The repulsion between lone pairs creates nano-compressive stresses that strengthen SGFs themselves to assist in raising the a_{uc} of the SGFRP-BMC system.

To summarize, the new double-step process of shortening SGFs using a 30 min extended mix, followed by 0.86 MGy HLEBI to finished samples, was found to increase the a_{uc} at the weak center of the SGFRP-BMC compression molded mother panel. This is caused by micro-strengthening by shortening fibers, as well as nano-strengthening by HLEBI.

However, for maximum safety, carefulness is highly recommended, since this study only applies to the panel center. HLEBI to the outer sections may weaken the composite. Also, it is always imperative to test for optimum dose of HLEBI for each situation. Nevertheless, the double-step process was found to increase impact values significantly, over 50%, at the weak center of the SGFRP-BMC mother panel.

4.4. Environmental Sustainability and Long-Term Durability Aspects

With the sharp increase in catastrophic events and heavy degradation to Earth's environment, environmental sustainability must be top priority in manufacture of any product. Therefore, we employ the double-step process to increase the impact strength of the SGFRP-BMC of the 30 min extended mix without a change in molding equipment, followed by HLEBI treatment without the use of any chemicals. However, for a full evaluation of the ecological footprint, life cycle assessment (LCA) is typically carried out. Energy consumption analysis of the double-step process would have to be taken into account. For the LCA to compare that with and without the double-step process would be needed. The LCA is evaluating the continuous cycle of raw materials, manufacturing, transportation, usage and selling, waste disposal, and recycling. The SGFRP-BMC is difficult to recycle, since the polymeric matrix is a thermoset that cannot be melted, for example, to separate it with filler and SGFs. Dumping SGFRP-BMC parts in a landfill is extremely hazardous for the environment and should be strictly prohibited. A highly employed remedy is to ground the waste GFRP-BMC and used as a filler [2]. Use of the filler recyclate for thermoplastic FRP would be recommended, since thermoplastics can be repeatedly melted and solidified for recyclability. As for long-term durability assessment, aging studies to check for mechanical strength reduction with time in treated SGFRP-BMC would be needed. LCA and aging studies are beyond the scope of this study, but should be considered for future research.

4.5. Economic Analysis, Scale-Up, and Feasibility Studies

Economic analysis, scale-up, and feasibility information are proprietary and beyond the scope of this study.

5. Conclusions

In order to maximize the safety of outdoor articles, airplanes, and space vehicles by protecting against the impact of airborne debris from increasing winds due to climate change, or from bird strikes or micrometeoroids, it is imperative for composite materials to have high impact resistance. In a 3-phase compression-molded short glass fiber polyester bulk-molded compound (SGFRP-BMC) with 55% wt. $CaCO_3$ filler and 11% wt. SGF, the center of the mother panel has lower impact strength than the outer sections, with solidification texture angles and SGF orientations being random from 0 to 90 degrees.

1. Therefore, a new double-step process of: (1) shortening the nominal 6.4 mm fiber length formulation to submilllimeter 0.44 mm by 30 min extended mixing without change in molding equipment, followed by (2) applying 0.86 MGy dose of homogeneous low-voltage electron beam irradiation (HLEBI) to both sides of the finished samples, requiring no chemicals or additives, which is shown to increase Charpy impact value (a_{uc}) over 50% from 6.26 to 9.59 kJm^{-2} at a median-accumulative probability of fracture, $P_f = 0.500$.
2. Shortening the SGFs by the extended mix method to submillimeter creates a higher number of thermal micro-compressive stress sites between SGF and the matrix to increase impact strength. This is performed by a mismatch of the coefficient of thermal expansion (CTE) between the matrix and fibers acting in the increased fiber spacing density while the composite is undergoing cool-down and shrinkage. In concert with this, since the SGFRP-BMC is a 3-phase fiber-filler-polymer system, the filled matrix behaves as a polymer-filler subsystem in the narrow spacing between fibers that efficiently allows for the increase in mechanical properties by decreasing the fiber length.
3. To boost impact strength further, HLEBI additionally provides nano-compressive stresses in the matrix by generating a dangling bond, which acts as repulsive force site between the outer-shell lone-pair electrons. This, along with increasing SGF/matrix adhesion, occurs with the optimum dose of HLEBI. During impact, a higher degree of internal cracking apparently occurs, raising the impact strength of SFGFRP-BMC samples.

Author Contributions: Conceptualization, M.C.F.; methodology, M.C.F. and Y.N.; software, M.C.F.; validation, M.C.F. and Y.N.; formal analysis, M.C.F.; investigation, M.C.F.; resources, M.C.F. and Y.N.; data curation, M.C.F.; writing—original draft preparation, M.C.F.; writing—review and editing, M.C.F. and Y.N.; visualization, M.C.F.; supervision, Y.N.; project administration, Y.N.; funding acquisition, Y.N. All authors have read and agreed to the published version of the manuscript.

Funding: This research received no external funding.

Institutional Review Board Statement: Not applicable.

Informed Consent Statement: Not applicable.

Data Availability Statement: Experimental data for this manuscript can be obtained upon request by contacting M.C. Faudree of Tokyo City University (faudree@tcu.ac.jp) or Yoshitake Nishi of Tokai University (west@tsc.u-tokai.ac.jp).

Acknowledgments: Shota Iizuka M.S., of Tokai University is acknowledged for his excellent assistance with HLEBI equipment. Sincere appreciation goes to Naoya Tsuchikura, Keisuke Iwata and Masae Kanda for their valuable input. Sincere gratitude goes to Citadel Plastics.

Conflicts of Interest: The authors declare no conflicts of interest.

Abbreviations

The following abbreviations are used in this manuscript.

SGFRP-BMC	short glass fiber reinforced polymer bulk molding compound
SGF	short glass fiber
CTE	coefficient of thermal expansion
HLEBI	homogeneous low-voltage electron beam irradiation
DB	dangling bond
FRP	fiber-reinforced polymer
Symbols	
a_{uc}	Charpy impact strength (kJm^{-2})
S_f	fiber spacing density (mm^{-3})
N_f	fiber number density (mm^{-3})
V_f	volume fraction
r	fiber radius (mm)
l_f	fiber length (mm)
BDE	bond dissociation energy (kJ mol^{-1})
D	HLEBI irradiation dose (MGy)
I	HLEBI irradiation current (mA)
S	HLEBI conveyor speed (mmin^{-1})
N	HLEBI number of irradiations
D_{th}	HLEBI penetration depth (μm)
V	HLEBI acceleration voltage (V)
ρ	sample density (g mm^{-3})
E	impact fracture energy (kJ)
W	impact hammer mass (kg)
R	length of hammer weight point from pivot center (m)
β	maximum angle after impact (Rad)
α	start angle before impact (Rad)
α'	maximum angle of a blank test (Rad)
b	specimen width (mm)
t	specimen thickness (mm)
P_f	accumulative probability
I_R	ascending strength rank
N_s	total number of samples in a data set

References

1. Faudree, M.; Nishi, Y.; Gruskiewicz, M.; Salvia, M. A new glass fibered reinforced composite with improved Charpy impact properties at low and high temperatures beyond the extremes of aircraft flight. *Mater. Trans.* **2018**, *59*, 1280–1287. [CrossRef]
2. Matykiewicz, D.; Barczewski, M.; Sterzyński, T. Morphology and thermomechanical properties of epoxy composites highly filled with waste bulk molding compounds (BMC). *J. Polym. Eng.* **2015**, *35*, 805–811. [CrossRef]
3. DeRosa, R.; Telfeyan, E.; Gaustad, G.; Mayes, S. Strength and microscopic investigation of unsaturated polyester BMC reinforced with SMC-Recyclate. *J. Thermoplast. Compos. Mater.* **2005**, *18*, 333–349. [CrossRef]
4. Rajaee, P.; Ghasemi, F.A.; Fasihi, M.; Saberian, M. Experimental Analysis and Optimization of Mechanical and Physical Properties of Light-Weight Bulk Molding Compound by Design of Experiment. *J. Macromol. Sci. Part B* **2021**, *60*, 237–256. [CrossRef]
5. Lee, T.; Jeong, K.; Kim, D. Development of a lightweight BMC material using fly ash. *Adv. Compos. Mater.* **2017**, *26*, 55–64. [CrossRef]
6. Faudree, M.; Nishi, Y.; Gruskiewicz, M. Characterization of velocity profile of highly-filled GFRP-BMC through rectangular-duct shaped specimen during injection molding from SEM fiber orientation mapping. *Mater. Trans.* **2013**, *54*, 1877–1883. [CrossRef]
7. Faudree, M.C.; Nishi, Y. Tensile Strength Enhancement by Shortening Glass Fibers with Sub-Millimeter Length in Bulk Molding Polymer Compound. *Mater. Trans.* **2010**, *51*, 2304–2310. [CrossRef]
8. Faudree, M.C.; Nishi, Y.; Gruskiewicz, M. A Novel 'Fiber Spacing' Model of Tensile Modulus Enhancement by Shortening Fibers to Sub-Millimeter in an Injection-Molded Glass Fiber Reinforced Polymer Bulk Molding Compound (GFRP-BMC). *Mater. Trans.* **2014**, *55*, 1292–1298. [CrossRef]
9. Singh, H.; Singh, T. Effect of fillers of various sizes on mechanical characterization of natural fiber polymer hybrid composites: A review. *Mater. Today Proc.* **2019**, *18*, 5345–5350. [CrossRef]
10. Rajaee, P.; Ghasemi, F.A.; Fasihi, M.; Saberian, M. Effect of styrene-butadiene rubber and fumed silica nano-filler on the microstructure and mechanical properties of glass fiber reinforced unsaturated polyester resin. *Compos. Part B Eng.* **2019**, *173*, 106803. [CrossRef]
11. Faudree, M.; Baer, E.; Hiltner, A.; Collister, J. Characterization of Damage and Fracture Processes in Short Fiber BMC Composites by Acoustic Emission. *J. Compos. Mater.* **1988**, *22*, 1170–1195. [CrossRef]
12. Tadlaoui, S.; Granger, R.; Vergnaud, J.M. Correlation Between Dynamic Mechanical Properties and State of Cure for Bulk Moulding Compounds. *Polym. Test.* **1994**, *13*, 271–284. [CrossRef]
13. Saburow, O.; Huether, J.; Maertens, R.; Trauth, A.; Kechaou, Y.; Henning, F.; Weidenmann, K.A. Adirect process to reuse dry fiber production waste for recycled carbon fiber bulk molding compounds. *Procedia CIRP* **2017**, *66*, 265–270. [CrossRef]
14. Lautenschläger, M.I.; Mayer, L.; Gebauer, J.; Weidenmanna, K.A.; Henning, F.; Elsner, P. Comparison of filler-dependent mechanical properties of jute fiber reinforced sheet and bulk molding compound. *Compos. Struct.* **2018**, *203*, 960–967. [CrossRef]
15. Sreenivasan, S.; Sulaiman, S.; Ariffin, M.K.A.M.; Baharudin, B.T.H.T.; Abdan, K. Physical Properties of Novel Kenaf Short Fiber Reinforced Bulk Molding Compounds (BMC) For Compression Moulding. *Mater. Today Proc.* **2018**, *5*, 1226–1232. [CrossRef]
16. Pang, J.W.C.; Fancey, K.S. Analysis of the tensile behaviour of viscoelastically prestressed polymeric matrix composites. *Compos. Sci. Technol.* **2008**, *68*, 1903–1910. [CrossRef]
17. Hadi, A.S.; Ashton, J.N. On the influence of pre-stress on the mechanical properties of a unidirectional GRE composite. *Compos. Struct.* **1997**, *40*, 305–311. [CrossRef]
18. Motahari, S.; Cameron, J. Fibre prestressed composites: Improvement of flexural properties through fibre prestressing. *J. Reinf. Plast. Compos.* **1999**, *18*, 279–288. [CrossRef]
19. Jenkins, C.; Donough, M.J.; Prusty, B.G. Mould free laminated composites using eccentric fibre prestressing. *Compos. Struct.* **2024**, *331*, 117867. [CrossRef]
20. Yuan, J.; Hiltner, A.; Baer, E.; Rahrig, R. The mechanical behaviour of PVC short-fibre composites. *J. Mater. Sci.* **1985**, *20*, 4377–4386. [CrossRef]
21. Meraghni, F.; Blakeman, C.J.; Benzeggagh, M.L. Effect of Interfacial Decohesion on Stiffness Reduction in a Random Discontinuous-Fibre Composite Containing Matrix Microcracks. *Compos. Sci. Technol.* **1996**, *56*, 541–555. [CrossRef]
22. Pittman, C.U.; Jiang, W.; Yue, Z.R.; Gardner, S.; Wang, L.; Toghiani, H.; Lyon y Leon, C.A. Surface properties of electrochemically oxidized carbon fibers. *Carbon* **1999**, *37*, 1797–1807. [CrossRef]
23. Hung, K.B.; Li, J.; Fan, Q.; Chen, Z.H. The enhancement of carbon fiber modified with electropolymer coating to the mechanical properties of epoxy resin composites. *Compos. Part A Appl. Sci. Manuf.* **2008**, *39*, 1133–1140. [CrossRef]
24. Tiwari, S.; Sharma, M.; Panier, S.; Mutel, B.; Mitschang, P.; Bijwe, J. Influence of cold remote nitrogen oxygen plasma treatment on carbon fabric and its composites with specialty polymers. *J. Mater. Sci.* **2011**, *46*, 964–974. [CrossRef]
25. Faudree, M.C.; Uchida, H.T.; Kimura, H.; Kaneko, S.; Salvia, M.; Nishi, Y. Advances in Titanium/Polymer Hybrid Joints by Carbon Fiber Plug Insert: Current Status and Review. *Materials* **2022**, *15*, 3220. [CrossRef] [PubMed]
26. Thomason, J.L.; Vlug, M.A. Influence on fibre length and concentration on the properties of glass fibre reinforced polypropylene: Tensile and flexural modulus. *Compos. Part A Appl. Sci. Manuf.* **1996**, *27*, 477–484. [CrossRef]
27. Thomason, J. The influence of fibre length, diameter and concentration on the modulus of glass fibre reinforced polyamide 6,6. *Compos. Part A Appl. Sci. Manuf.* **2008**, *39*, 1732–1738. [CrossRef]
28. Thomason, J.L.; Vlug, M.A. Influence on fibre length and concentration on the properties of glass-fibre-reinforced polypropylene: Impact properties. *J. Compos. Part A* **1997**, *28*, 277–278. [CrossRef]

29. Fu, S.-Y.; Lauke, B. Effects of fiber length and fiber orientation distributions on the tensile strength of short fiber reinforced polymers. *Compos. Sci. Technol.* **1996**, *56*, 1179–1190. [CrossRef]
30. Huang, H.; Talreja, R. Numerical simulation of matrix micro-cracking in short fiber reinforced polymer composites: Initiation and propagation. *Compos. Sci. Technol.* **2006**, *66*, 2743–2757. [CrossRef]
31. Thomason, J.L. The influence of fibre properties of the performance of glass fiber reinforced polyamide 6,6. *Compos. Sci. Technol.* **1999**, *59*, 2315–2328. [CrossRef]
32. Rezaei, F.; Yunus, R.; Ibrahim, N.A.; Mahdi, E.S. Effect of fiber loading and fiber length on mechanical and thermal properties of short carbon fiber reinforced polypropylene composite. *Malays. J. Anal. Sci.* **2007**, *11*, 181–188.
33. Capela, C.; Oliveira, S.; Pestana, J.; Ferreira, J. Effect of fiber length on the mechanical properties of high dosage carbon reinforced. *Procedia Struct. Integr.* **2017**, *5*, 539–546. [CrossRef]
34. Haghighatnia, T.; Abbasian, A.; Morshediane, J. Hemp fiber reinforced thermoplastic polyurethane composite: An investigation in mechanical properties. *Ind. Crops Prod.* **2017**, *108*, 853–863. [CrossRef]
35. Maertens, R.; Hees, A.; Schöttl, L.; Liebig, W.; Elsner, P.; Weidenmann, K.A. Fiber shortening during injection molding of glass fiber-reinforced phenolic molding compounds: Fiber length measurement method development and validation. *Polym. Technol. Mater.* **2021**, *60*, 872–885. [CrossRef]
36. Faudree, M.C.; Nishi, Y.; Salvia, M. Increasing Impact Strength of a Short Glass Fiber Compression Molded BMC by Shortening Fibers without Change in Equipment. *Materials* **2022**, *15*, 1145. [CrossRef] [PubMed]
37. Panaitescu, D.M.; Fierascu, R.C.; Gabor, A.R.; Nicolae, C.A. Effect of hemp fiber length the mechanical and thermal properties of polypropylene/SEBS/hemp fiber composites. *J. Mater. Res. Technol.* **2020**, *9*, 10768–10781. [CrossRef]
38. Senthilrajan, S.; Venkateshwaran, N.; Naresh, K.; Velmurugan, R.; Gupta, N.K. Effects of jute fiber length and weight percentage on quasi-static flexural and dynamic mechanical properties of jute/polyester composites for thin-walled structure applications. *Thin-Walled Struct.* **2022**, *179*, 109719.
39. Sajin, J.B.; Christu Paul, R.; Binoj, J.S.; Brailson Mansingh, B.; Gerald Arul Selvan, M.; Goh, K.L.; Rimal Isaac, R.S.; Senthil Saravanan, M.S. Impact of fiber length on mechanical, morphological and thermal analysis of chemical treated jute fiber polymer composites for sustainable applications. *Curr. Res. Green Sustain. Chem.* **2022**, *5*, 10024. [CrossRef]
40. Bhagat, A.B.; Ghosh, A.K. Estimation of rheological percolation threshold and influence of fibre length on properties of polypropylene/sisal fibre composites having near critical fibre length. *Polymer* **2022**, *258*, 125304. [CrossRef]
41. Aravinth, K.; Sathish, R.; Ramakrishnan, T.; Balu Mahandiran, S.; Shiyam Sundhar, S. Mechanical investigation of agave fiber reinforced composites based on fiber orientation, fiber length, and fiber volume fraction. *Mater. Today Proc.* 2024, *in press*. [CrossRef]
42. Starink, M.; Syngellakis, S. Shear lag models for discontinuous composites: Fibre end stresses and weak interface layers. *Mater. Sci. Eng. A* **1999**, *270*, 270–277. [CrossRef]
43. Bauccio, M.L. (Ed.) *ASM Engineered Materials Reference Book*, 2nd ed.; ASM International: Novelty, OH, USA, 1994.
44. Calcium Carbonate, Calcite (CaCO3). Available online: https://www.matweb.com/search/datasheet_print.aspx?matguid=bea4bfa9c8bd462093d50da5eebe78ac (accessed on 23 February 2024).
45. Dragoi, D.; Üstündag, E.; Clausen, B.; Bourke, M.A. Investigation of thermal residual stresses in tungsten-fiber/bulk metallic glass matrix composites. *Scr. Mater.* **2001**, *45*, 245–252. [CrossRef]
46. Seol, K.; Krawitz, A.; Richardson, J.; Weisbrook, C. Effects of WC size and amount on the thermal residual stress in WC–Ni composites. *Mater. Sci. Eng. A* **2005**, *398*, 15–21. [CrossRef]
47. Niihara, K. Overcoming the fragility of ceramics (Challenge to strengthen ceramics). *Ceram. Kyoujinka Fract. Toughness Ceram.* **1986**, *21*, 581–589. (In Japanese)
48. Wu, C.; Shen, S.; Li, Y.; Luo, G.; Shen, Q.; Gan, Z.; Liu, J. Influence of coarse grain particles on mechanical properties and fracture behavior in multi-modal Al-based metal matrix composites. *Powder Technol.* **2021**, *394*, 901–908. [CrossRef]
49. Faudree, M.; Nishi, Y.; Gruskiewicz, M. Effects of electron beam irradiation on Charpy impact value of short glass fiber (GFRP) samples with random distribution of solidification texture angles from zero to 90 degrees. *Mater. Trans.* **2012**, *53*, 1412–1419. [CrossRef]
50. Nishi, Y.; Kobayashi, H.; Salvia, M. Effects of Electron Beam Irradiation on Charpy Impact Value of GFRP. *Mater. Trans.* **2007**, *48*, 1924–1927. [CrossRef]
51. James, A.; Lord, M. *Macmillan's Chemical and Physical Data, London and Basingstoke*; The Macmillan Press, Ltd.: London and Basingstoke, UK, 1992; pp. 484–485.
52. Gordon, A.; Ford, R. *The Chemists Companion—A handbook of Practical Data, Techniques, and References*; A Wiley Interscience Publication: New York, NY, USA, 1972; pp. 112–113.
53. *ASTM D 6110-02*; Standard Test Methods for Determining the Charpy Impact Resistance of Notched Specimens of Plastics. American Society for Testing and Materials: West Conshohocken, PA, USA, 2002.
54. Christenhusz, R.; Reimer, L. Schichtdickenabhangigkeit der warmerzeugungdurch elektronenbestrahlung im energiebereich zwischen 9 und 100 keV (Layer thickness dependency of heat generation by electron irradiation in the energy range between 9 and 100 keV). *Z. Angew. Phys.* **1967**, *23*, 396–404.
55. *JIS K 7077*; Testing Method for Charpy Impact Strength of Carbon Fiber Reinforced Plastics. Japanese Industrial Standards Committee: Tokyo, Japan, 1991. (In Japanese)

56. Splett, J.; Iyer, H.; Wang, C.; McCowan, C. *National Institute of Standards and Technology (NIST) Recommended Practice Guide, Computing Uncertainty for Charpy Impact Test Machine Test Results*; Special publication 960-18; US Department of Commerce: Boulder, Colorado, 2008; pp. 27–29.
57. Nishida, T.; Yasuda, E. *Evaluation of Dynamic Properties of Ceramics (in Japanese, Ceramics no rikigaku tokusei hyouka)*; Nikkan Kogaku Shimbun Sha: Tokyo, Japan, 1986; pp. 50–51.

materials MDPI

Communication

Polyacrylonitrile Composites Blended with Asphalt as a Low-Cost Material for Producing Synthetic Fibers: Rheology and Thermal Stability

Artem V. Pripakhaylo [1,*], Alexei A. Tsypakin [2], Anton A. Klam [2], Andrei L. Andreichev [2], Andrei R. Timerbaev [1], Oksana V. Shapovalova [1] and Rustam N. Magomedov [1,*]

[1] Semenov Research Center for Chemical Physics, Russian Academy of Sciences, Kosygin St. 4, 119991 Moscow, Russia; andrei.timerbaev@mail.ru (A.R.T.); shapovalova@chph.ras.ru (O.V.S.)
[2] UMATEX, Varshavskoe Sh. 46, 115230 Moscow, Russia; a.tsipakin@umatex.ru (A.A.T.)
* Correspondence: artem.pripakhaylo@chph.ras.ru (A.V.P.); rustam.magomedov@chph.ras.ru (R.N.M.)

Abstract: The results of rheological studies and thermal analysis of polymer compositions based on polyacrylonitrile copolymers (PAN) of different molecular weights and asphalt isolated by n-pentane solvent deasphalting are presented. It was found that the asphalt content in mixtures with PAN at the level of 10–30 wt.% improves the rheological properties of the polymer composite melt. In particular, the temperatures of extrusion and molding of fibers tend to reduce, and the time during which the melt retains its rheological characteristics necessary for extrusion is notably increased, from 43 to 92 min. Thermal analysis by DSC revealed no effect of asphalt additive in an amount of up to 30 wt.% on radical PAN cyclization and the subsequent stage of fiber stabilization. Our study proved the possibility of preparing polymer composites based on PAN and asphalt suitable for extrusion and eventual molding of continuous filaments of synthetic fibers with reduced cost of production.

Keywords: polyacrylonitrile; asphalt; rheological properties; plasticizing additive

1. Introduction

The main type of raw materials from which the production of textile and high-strength fibers is carried out is polyacrylonitrile copolymers (PAN). The PAN fibers are generally produced via an expensive solution-molding technology, which is also environmentally unfriendly due to the use of large amounts of toxic organic solvents [1]. On the contrary, fiber formation from PAN melt needs no solvents and eliminates polymer losses upon increasing the production speed.

Nowadays, more than in years before, there is a growing concern of polymer chemists to reduce the production cost of synthetic fibers, including those based on PAN, which is widely used both in the textile industry and in the production of reinforcing materials. Perhaps the most straightforward approach is employing PAN blends with cheaper components. For instance, for the production of low-cost fibers, Ding et al. proposed to employ lignin in the amount of 10–45 wt.% of the total weight of PAN [2,3]. The resulting composite fibers are characterized by increased fire resistance, while carbon fibers (CFs) possess high tensile strength. However, neither PAN fibers blended with lignin nor the respective CFs can reach the mechanical properties of parental PAN fibers [4].

Another class of promising additives includes a cheap by-product of solvent deasphalting (SDA), asphalt, characterized by high softening temperatures [5–7]. The conversion of SDA products into the CF directly, without any energy-intensive preparation steps, has been the focus of research by Saad et al. [8]. and Leistenschneider and coworkers [9]. Asphalts of pentane (C5) SDA and C5-asphaltenes were used in these studies as inexpensive feedstocks for producing the CF. It is worth noting that n-pentane provides the final product as dry solid granules or powder [10]. In addition, the resulting C5-asphaltenes feature a high softening temperature (more than 170 °C). On the other hand, to boost the physical and mechanical properties of CF, it is required to treat a pitch precursor at the stage of stabilization with HNO_3 followed by heat treatment [11]. The obtained CFs are distinct in an elastic modulus of 71 GPa and tensile strength of 1130 MPa, which, however, are inferior to similar parameters of CFs based on an individual PAN. Thus, asphalt is an additive to PAN that holds the promise of achieving satisfactory physical and mechanical indicators of CFs while reducing the total cost of their production.

However, the polymeric compositions with different additive contents may not be easily subject to subsequent fiber formation. One of the main methods for the preliminary assessment of the melt processability/spinnability is rheology research. The melt-spinning ability is determined by the thermal rheological behavior under theoretical forming conditions. The temperature-dependent complex viscosity profile, as well as the accumulation and loss modulus (G' and G'', respectively), are useful factors to evaluate the minimum values of process parameters that a PAN composition can achieve before the cross-linking between the nitrile groups reaches critical values [12,13].

Our work is aimed at studying the rheological and thermal behaviors of PAN-based compositions blended with asphalt and assessing their prospective for molding the polymer fibers by the spin-melt method. The results described below give impetus to developing the theoretical basis of the molding of plasticized melts intended to produce cheap synthetic fibers based on PAN.

2. Materials and Methods

2.1. Materials

The PAN polymers of the same composition but different molecular weights, 55 kDa (PAN-1) and 91 kDa (PAN-2), were obtained via radical (precipitation) copolymerization. Differing significantly in molecular weight, the synthesized PAN copolymers feature different rheological properties. PAN-1 becomes viscous when heated to 172 °C and can therefore be potentially melt-formed into fiber. For PAN-2, the transition to a viscous state is possible only for a short period of time at 184 °C, which does not allow the formation of the respective fiber. As a plasticizer, the asphalt of C5-SDA of vacuum residue was used, with a C7-asphaltene content of 52 wt.%.

2.2. C5-SDA Procedure

Asphalt was produced in a high-pressure autoclave with a volume of 2 L, equipped with an anchor mixer and an external electric heating jacket. An electric belt heater heated a drain valve installed at the bottom of the vessel to maintain the minimum permissible viscosity of the asphalt during sampling. After the vacuum residue sample (oil refinery, Russia) was cooled to room temperature, heavy oil feedstock and n-pentane (99.0% wt, Ekos-1 JSC, Moscow, Russia) were added to the autoclave to avoid solvent evaporation. The autoclave was sealed, heated to 170 °C, and stirred. The agitator's rotation speed during extraction was 600 rpm. If necessary, after reaching the operating extraction temperature, the pressure was slowly elevated to 50 bar by adding additional amounts of solvent using a plunger pump. The total ratio of n-pentane to the vacuum residue sample was 6/1 (v/v). The extraction time was counted from the moment the set values of temperature, pressure, and rotation speed of the agitator were established. The extraction process was completed after 30 min, and the mixture was kept without stirring for 30 min. Upon phase separation, the asphalt was removed through a bottom valve and brought to a constant mass in a drying oven at a temperature not exceeding 120 °C [14]. The content of asphaltenes insoluble in n-heptane was determined according to the IP 143 (ASTM D 6560) method [15].

2.3. Polymer Blend Preparation

Composite samples were prepared by mechanical grinding and mixing of solid PAN and asphalt to form a homogeneous dispersed product using a porcelain mortar and pestle. Prior to mixing, PAN and asphalt were dried in a drying oven at a temperature of 100 °C until a constant mass. Samples were molded under pressure of about 5 tons to produce a round tablet with a diameter of 1.5 cm and a thickness of 1.2–1.3 mm. Figure 1 shows a comparison of polymer composites obtained by grinding and pressing.

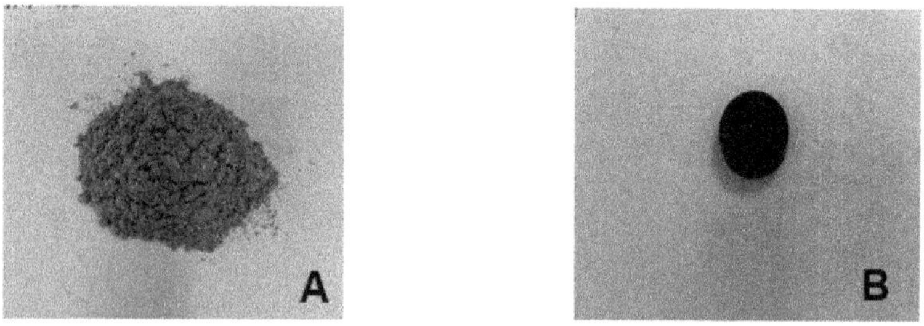

Figure 1. Images of the polymer composite after (**A**) grinding and (**B**) pressing.

2.4. Polymer Characterization

The molecular weight of polymers was determined by gel-permeation chromatography according to ISO 16014-1:2019 [16] and ISO 16014-3:2019 [17] on a Shimadzu Prominence LC-20 instrument (Shimadzu, Kyoto, Japan) equipped with a RID-20A differential refractometer. The elution was performed with DMF containing 1 wt.% LiBr at 60 °C. DSC analyses were performed by ISO 11357-5:2013 [18] using a DSC 214 Polyma analyzer (Netchzt, Hanau, Germany) with heating from 50 to 400 °C at a rate of 10 °C/min. in the air atmosphere. A modular rheometer Anton Paar MSR 102 (Anton Paar GmbH, Graz, Austria), operating in the range from 140 to 190 °C, was used for rheological studies.

Temperature dependencies of the storage and loss modules were recorded in two consecutive modes. According to mode (1), the sample was heated at a rate of 2 °C/min in the range of 165 to 190 °C so that the temperature of the phase transition from the elastic to the viscoelastic state (T_1) can be determined from the equality of G' and G'' at the tangent of the mechanical loss angle (tg δ) = 1. Upon heating, the moduli become equal when the material transitions from an elastic to viscoelastic state. Therefore, T_1 can be considered as the melting point of the material. As the material is further heated, tg δ increases and the next significant point is tg δ = 1.2. It is reached at a certain temperature value ($T_{1.2}$ thereafter) when the material becomes suitable for melt spinning and obtaining continuous filaments from the melt. In mode (2), the sample was subjected to isothermal heating at $T_{1.2}$ to estimate the melt lifetime, i.e., the time at which the material is characterized by tg $\delta \geq 1.2$.

3. Results and Discussion

3.1. Thermal Analysis

Of utmost interest in the DSC thermograms of polymers is the characteristic exothermic peak in the region of 220–300 °C [19,20], which portrays the complex reactions of thermal oxidation, as well as intra- and intermolecular cyclization (mainly by CN groups). The latter is the basis for the thermal stabilization of PAN fibers and their further high-temperature processing. As can be seen in Figure 2, the DSC peak of initial PAN has a characteristic temperature range (220–400 °C), with a maximum at about 320 °C, and asphalt additives from 10 to 30 wt.% exert no significant effect on its shape and position along the temperature scale.

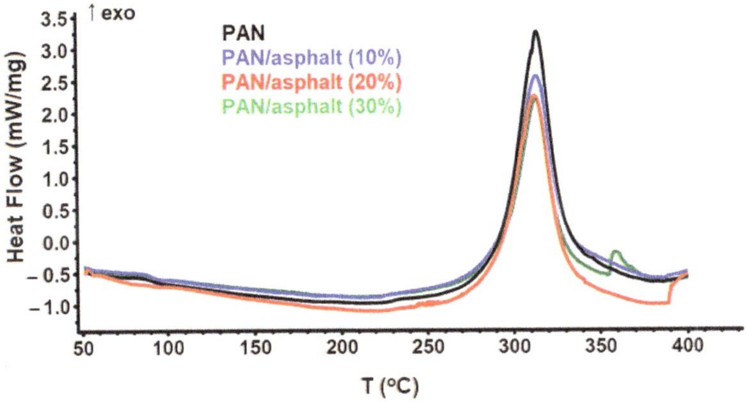

Figure 2. DSC-thermograms of the initial PAN and polymer composites with asphalt.

The results of the DSC analysis are presented in Table 1. It should be noted that the decrease in the peak area (or reaction enthalpy) for PAN–asphalt mixtures is probably due to the endothermic effect of the cracking reaction of petroleum components [21,22]. The thermogram of the original asphalt is characterized by an endothermic effect at temperatures above 370 °C associated with the cracking of petroleum asphaltenes, which is confirmed in the literature [23,24].

Table 1. Results of DSC analysis of initial PAN-1 and polymer composites with asphalt.

Sample	Peak Onset Temperature [°C]	Peak Temperature [°C]	Peak Area [J/g]
PAN-1	227.9	331.4	623.8
PAN-1 + 10% asphalt	227.5	312.5	547.4
PAN-1 + 20% asphalt	225.8	313.6	523.7
PAN-1 + 30% asphalt	224.2	313.3	476.6

3.2. Rheological Studies

For these studies, we employed the oscillation test mode with controlled shear deformation and constant angular velocity. The advantage of such testing is the ability of measurements without destroying the sample structure and the detailing of rheological characteristics by isolating elastic and viscous components. The characteristics in question were the storage modulus G' (elastic component), the loss modulus G'' (viscous component), and their ratio, i.e., the tangent of the mechanical loss angle tg δ, determined at a strain value of 1%, a thickness of 1 mm, and a constant angular velocity of 31.5 rad/s.

For asphalt, the rheological properties were assessed in the range of 140–190 °C. The lower limit of the temperature range was selected experimentally via a gradual decrease in the initial temperature until the temperature of the intersection of the G' and G'' curves was attained. It is important to note that at temperatures close to and above 190 °C, PAN-based samples are subject to a release of gaseous products due to oxidation and cyclization processes [25]. Such a phenomenon seriously distorts the rheological properties and makes the processes of extrusion and spinning technologically impossible (let alone a comparison between PAN and asphalt). Therefore, we avoided rheological measurements beyond the specified temperature range.

Rheological curves defining the temperature behavior of G' and G'' are shown in Figure 3.

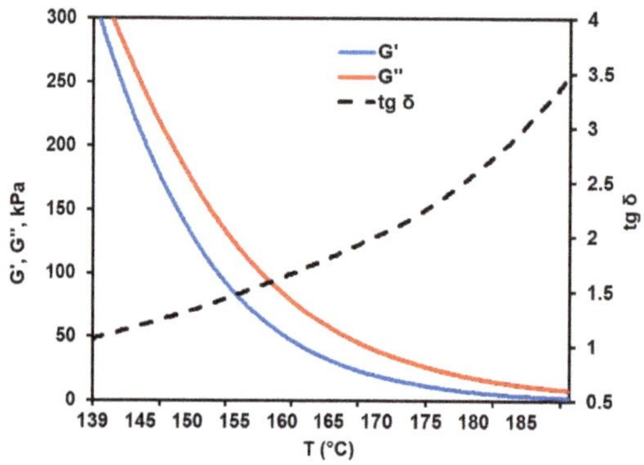

Figure 3. Rheological curves of asphalt.

Throughout the entire temperature range, the G″ exceeds the G′, which is indicative of the viscous state of asphalt. At the same time, both G″ and G′ values decrease with increasing temperature, while the G″/G′ ratio gradually increases from 1.05 at 140 °C to 2.58 at 180 °C. These observations imply the absence of thermal cross-linking, destruction, and other adverse temperature effects on the rheological properties in the range under scrutiny [26]. At temperatures between 170 and 180 °C, the complex viscosity of asphalt reduces from 1288 to 569 Pa × s, remaining below the same parameter of the original PAN samples (see Table 2). This opens the opportunity of using asphalt as a plasticizing additive.

Table 2. Viscosity data.

Sample	Complex Viscosity [Pa × s]	
	170 °C	180 °C
Asphalt	1288	569
PAN-1	1487	970
PAN-2	2620	1980

Figure 4 shows rheological curves for the PAN-1 and PAN-2 polymers prepared as described in the Materials and Methods section. The minimum temperature was chosen as 165 °C because at lower temperatures the samples are in a solid state, which prevents them from gaining the dimensions necessary for testing. The rheological characteristics of PAN-1 and PAN-2 are fairly different. Specifically, the intersection point for G′ and G″, corresponding to the transition from an elastic to a viscous state [12], is reached at 172 and 184 °C for PAN-1 and PAN-2, respectively. The critical value of tg δ, i.e., 1.2, is observed for PAN-1 at 176 °C ($T_{1.2}$), at which the melt displays a "lifetime" of 43 min. In contrast, for PAN-2, this value is not achievable over the entire range studied, and the lifetime has not been determined. Our preliminary studies have shown that at tg $\delta \leq 1.2$, PAN melts cannot form continuous fibers, although the extrusion process remains an opportunity.

For polymer composites based on PAN-1, rheological curves are presented in Figure 5. The temperature of the point of intersection, as well as $T_{1.2}$, tends to shift toward lower values with an increase in asphalt content, thereby confirming its plasticizing action. A similar effect of asphalt is also evident from the data on the lifetime of polymer compositions (Table 3).

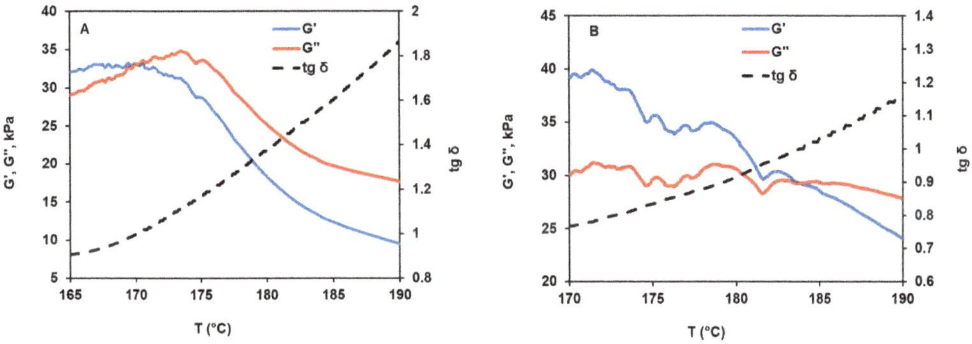

Figure 4. Rheological curves of PAN-1 (**A**) and PAN-2 (**B**).

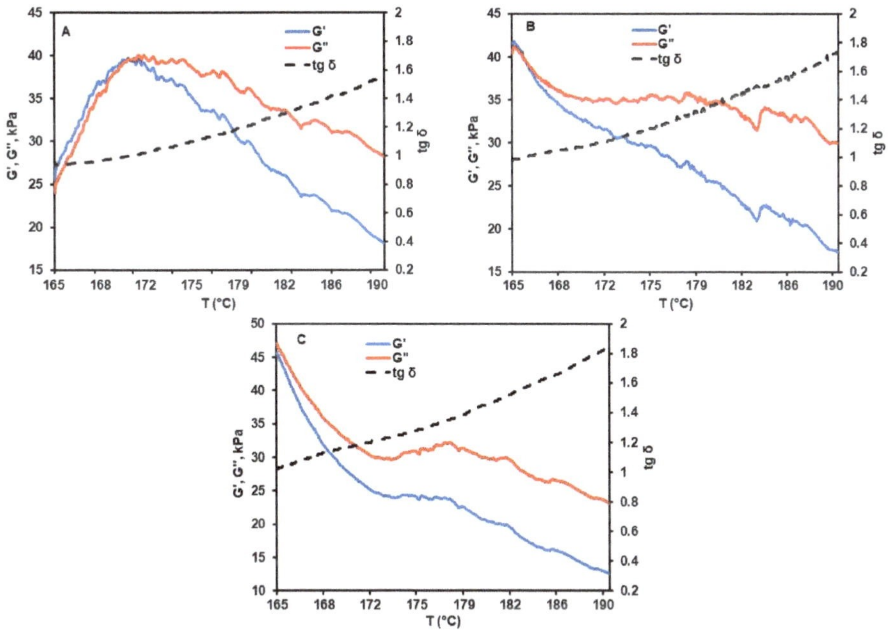

Figure 5. Rheological curves of PAN-1 composites with (**A**)—10; (**B**)—20; and (**C**)—30 wt.% asphalt.

Table 3. Rheological parameters of PAN-1 composites.

Sample	T_1 [°C] [a]	$T_{1.2}$ [°C]	Lifetime at $T_{1.2}$ [min]	Complex Viscosity at $T_{1.2}$ [Pa × s]
PAN-1	172	176	43	1363
PAN-1 + 10% asphalt	170	175	56	1229
PAN-1 + 20% asphalt	167	173	87	1431
PAN-1 + 30% asphalt	164	172	92	1390

[a] At tg δ = 1.

For example, the addition of 20 wt.% of asphalt leads to a more than two-fold increase in the lifetime, up to 87 min. However, a further increase in asphalt content has no significant effect. Apparently, longer lifetimes of the melt are mainly associated with decreased $T_{1.2}$ and deceleration of the cyclization processes by CN groups. This enables the melt to retain its rheological properties and fiber-formation potential for a longer time. It should be mentioned that the complex viscosity of melts with different asphalt contents does not notably differ from the viscosity of the original PAN-1 melt. Therefore, when lowering the melt viscosity is an issue, using asphalt as an additive makes an untoward choice.

To further assess the prospect of using asphalt for blending PAN-based melts, we evaluated the rheological parameters for higher molecular weight PAN-2. For the corresponding polymer compositions, similar rheological curves were recorded. Since PAN-2 has a softening point higher than PAN-1 (due to its higher molecular weight), the initial temperature was fixed at 170 °C, while the maximum temperature remained the same, 190 °C. The rheological data obtained are listed in Table 4.

Table 4. Rheological parameters of PAN-2 composites.

Sample	T_1 [°C] [a]	$T_{1.2}$ [°C]	Lifetime at $T_{1.2}$ [min]	Complex Viscosity at $T_{1.2}$ [Pa × s]
PAN-2	184	-	-	-
PAN-2 + 10% asphalt	181	188	3.5	1520
PAN-2 + 20% asphalt	171	180	9.0	2128
PAN-2 + 30% asphalt	169	178	15.3	2316

[a] At tg $\delta = 1$.

Obviously, asphalt additives to PAN-2 favor the melt lifetime, holding the trend of gradually reducing T_1 and $T_{1.2}$. However, this effect seems insufficient for utilizing these composites for spin melting. Even for the melt fortified with 30 wt.% of asphalt, the lifetime is much shorter than 45 min, which is the minimum time required to maintain the extrusion and molding processes uninterrupted.

4. Conclusions

The principal possibility of using asphalt obtained by SDA of vacuum residue as an external plasticizer of PAN melts is demonstrated. In the selected temperature region, asphalt does not adversely affect the rheological properties of the polymer material and reduces the temperature of transition to a viscous state and eventually, the temperature of fiber formation of polymer composites. This extends the lifetime of the melt and should improve its moldability. Another anticipated advantage of asphalt compared to other plasticizers (ionic liquids, glycols, propylene carbonate, etc.) is that it does not need removal after the fiber is obtained. However, the main benefit of the proposed blending approach is perhaps due to the low cost of oil residues and the SDA process itself, which would inevitably reduce the total cost of production of PAN-based fibers. Our ongoing research is directed to the optimization of composition and rheological properties of PAN–asphalt composites, as well as to achieving insight into the extrusion, fibrillation, and drawing of fibers based on the optimum blends and the assessment of their applicability in various industries (automotive, civil engineering, production of sporting goods and various equipment, etc.).

Author Contributions: Conceptualization, A.V.P., A.A.T. and R.N.M.; writing—original draft, A.V.P., A.A.T. and A.A.K.; formal analysis, A.V.P., A.A.T. and A.A.K.; writing—review and editing, A.V.P., R.N.M., A.R.T. and A.A.T.; visualization, A.A.T. and A.A.K.; supervision, R.N.M., A.L.A. and A.R.T.; investigation, A.A.T. and A.A.K.; funding acquisition, R.N.M. and O.V.S.; project administration, O.V.S., A.L.A. and R.N.M. All authors have read and agreed to the published version of the manuscript.

Funding: The work was performed within the framework of the state assignment of the Semenov Research Center for Chemical Physics (No. 1023110900097-3-2.4.2).

Institutional Review Board Statement: Not applicable.

Informed Consent Statement: Not applicable.

Data Availability Statement: The original contributions presented in the study are included in the article, further inquiries can be directed to the corresponding author.

Conflicts of Interest: Authors Alexei A. Tsypakin, Anton A. Klam, Andrei L. Andreichev were employed by the UMATEX. The remaining authors declare that the research was conducted in the absence of any commercial or financial relationships that could be construed as a potential conflict of interest.

References

1. Peijs, T.; Kirschbaum, R.; Lemstra, P.J. A critical review of carbon fiber and related products from an industrial perspective. *Adv. Ind. Eng. Pol. Res.* **2022**, *5*, 90–106. [CrossRef]
2. Ding, R.; Wu, H.; Thunga, M.; Bowler, N.; Kessler, M.R. Processing and characterization of low-cost electrospun carbon fibers from organosolv lignin/polyacrylonitrile blends. *Carbon* **2016**, *100*, 126–136. [CrossRef]
3. Bissett, P.J.; Herriott, C.W. Lignin/Polyacrylonitrile-Containing Dopes, Fibers, and Methods of Making Same. U.S. Patent No. 9133568, 15 September 2015.
4. Souto, F.; Calado, V.; Pereira, N. Lignin-based carbon fiber: A current overview. *Mater. Res. Exper.* **2018**, *5*, 072001. [CrossRef]
5. Bisheh, H.; Abdin, Y. Carbon fibers: From PAN to asphaltene precursors. A state-of-art review. *J. Carbon Res.* **2023**, *9*, 19. [CrossRef]
6. Al Bari, M.A.; Nabil, S.K.; Saad, S.; Sarkar, R.; Sabiha, S.; Rahman, M.M.; Kibria, M.G. Economic and environmental assessment of asphaltene-derived carbon fiber production. *Green Chem.* **2023**, *24*, 6446–6458. [CrossRef]
7. Karaaslan, M.A.; Gunning, D.; Huang, Z.; Ko, F.; Renneckar, S.; Abdin, Y. Carbon Fibers from Bitumen-Derived Asphaltenes: Strategies for Optimizing Melt Spinnability and Improving Mechanical Properties. *Carbon* **2024**, *228*, 119300. [CrossRef]
8. Saad, S.; Zeraati, A.S.; Roy, S.; Saadi, M.A.S.R.; Radović, J.R.; Rajeev, A.; Kibria, M.G. Transformation of petroleum asphaltenes to carbon fibers. *Carbon* **2022**, *190*, 92–103. [CrossRef]
9. Leistenschneider, D.; Zuo, P.; Kim, Y.; Abedi, Z.; Ivey, D.G.; de Klerk, A.; Chen, W. A mechanism study of acid-assisted oxidative stabilization of asphaltene-derived carbon fibers. *Carbon Trends* **2021**, *5*, 100090. [CrossRef]
10. Shi, Q.; Zhao, S.; Zhou, Y.; Gao, J.; Xu, C. Development of heavy oil upgrading technologies in China. *Rev. Chem. Eng.* **2019**, *36*, 1–19. [CrossRef]
11. Zuo, P.; Leistenschneider, D.; Kim, Y.; Abedi, Z.; Ivey, D.G.; Zhang, X.; Chen, W. Asphaltene thermal treatment and optimization of oxidation conditions of low-cost asphaltene-derived carbon fibers. *J. Ind. Eng. Chem.* **2021**, *104*, 427–436. [CrossRef]
12. Jiang, J.; Srinivas, K.; Kiziltas, A.; Geda, A.; Ahring, B.K. Rheology of polyacrylonitrile/lignin blends in ionic liquids under melt spinning conditions. *Molecules* **2019**, *24*, 2650. [CrossRef] [PubMed]
13. Tian, Y.C.; Han, K.Q.; Qin, H.L.; Rong, H.P.; Yan, B.; Wang, D.; Yu, M.H. Rheological behaviors of polyacrylonitrile melt using ionic liquids as a plasticizer. *Adv. Mater. Res.* **2012**, *476*, 2151–2157.
14. Magomedov, R.N.; Pripakhailo, A.V.; Maryutina, T.A. Effect of the phase state of the solvent on solvent deactivation of tar by n-pentane. *Chem. Technol. Fuels Oils* **2019**, *54*, 721–732. [CrossRef]
15. *ASTM D 6560*; Standard Test Method for Determination of Asphaltenes (Heptane Insolubles) in Crude Petroleum and Petroleum Products. ASTM International: West Conshohocken, PA, USA, 2022.
16. *ISO 16014-1:2019*; Plastics—Determination of average molecular mass and molecular mass distribution of polymers using size-exclusion chromatography. International Organization for Standardization: Geneva, Switzerland, 2019.
17. *ISO 16014-3:2019*; Plastics—Determination of average molecular weight and molecular weight distribution of polymers using size-exclusion chromatography. International Organization for Standardization: Geneva, Switzerland, 2019.
18. *ISO 11357-5:2013*; Plastics—Differential scanning calorimetry (DSC). Part 5: Determination of characteristic reaction-curve temperatures and times, enthalpy of reaction and degree of conversion. International Organization for Standardization: Geneva, Switzerland, 2013.
19. Morgan, P. *Carbon Fibers and Their Composites*; Taylor & Francis Group: Boca Raton, LA, USA, 2005; p. 192.
20. Morris, E.A.; Weisenberger, M.C. High performance carbon fibers from very high molecular weight polyacrylonitrile precursors. *Carbon* **2016**, *101*, 245–252. [CrossRef]
21. Ebrahimi, S.; Moghaddas, J.S.; Aghjeh, M.K.R. Study on thermal cracking behavior of petroleum residue. *Fuel* **2008**, *87*, 623–1627. [CrossRef]
22. Al-Humaidan, F.; Hauser, A.; Al-Rabiah, H.; Lababidi, H.H.; Bouresli, R. Studies on thermal cracking behavior of vacuum residues in Eureka process. *Fuel* **2013**, *109*, 635–646. [CrossRef]
23. Yasar, M.; Akmaz, S.; Gurkaynak, M.A. Investigation of glass transition temperatures of Turkish asphaltenes. *Fuel* **2007**, *86*, 1737–1748. [CrossRef]
24. Korneev, D.S.; Pevneva, G.S.; Golovko, A.K. Study of the thermal stability of petroleum asphaltenes by differential scanning calorimetry. *AIP Conf. Proc.* **2018**, *2051*, 020134-1–020134-4.

25. Liu, W.; Cheng, L.; Zhang, H.; Zhang, Y.; Wang, H.; Yu, M. Rheological behaviors of polyacrylonitrile/1-butyl-3-methylimidazolium chloride concentrated solutions. *Int. J. Mol. Sci.* **2007**, *8*, 180–188. [CrossRef]
26. Eshraghian, A.; Kamkar, M.; Sundararaj, U. Asphaltene/polymer composites: Morphology, compatibility, and rheological properties. *Can. J. Chem. Eng.* **2023**, *101*, 1421–1439. [CrossRef]

materials

Article

Hydrogels Based on Polyacrylamide and Pectin Containing Rice Husk Ash: Preparation, Characterization and Application in Formulation of Cementitious Materials

Ruth Hevellen Sousa Rodrigues [1], Edson Araujo de Almeida [2], Fábio Rodrigo Kruger [3], Edson Cavalcanti Silva-Filho [1] and Edvani Curti Muniz [1,2,4,*]

1 Chemistry Postgraduation Program, Federal University of Piaui—UFPI, Teresina 64049-550, PI, Brazil
2 Chemistry Postgraduation Program, State University of Maringa—UEM, Maringá 87020-900, PR, Brazil
3 Civil Construction Department, Federal University of Technology-Parana—UTFPR-CM, Campo Mourão 87301-889, PR, Brazil
4 Materials Science Postgraduation Program, Federal University of Technology-Parana—UTFPR-LD, Londrina 86036-370, PR, Brazil
* Correspondence: munizec@ufpi.edu.br

Abstract: Superabsorbent polymers (PSAs) have been extensively studied to act as internal curing agents in cementitious materials, as they have the characteristic of absorbing and releasing water in a controlled manner, which can contribute to the hydration process of a cementitious medium during its consolidation. Thus, hydrogels consisting of polyacrylamide (PAAm), pectin (Pec) and rice husk ash (RHA) were synthesized to be applied in cementitious matrices. In addition, the PSAs were characterized by FTIR, SEM, and XRD. For evaluating the usage of hydrogels as internal curing agents, different hydrogel contents—0.03, 0.06, and 0.1 (wt-%, relative to cementitious components)—were used for mortar preparation. The mechanical strengths of the cementitious materials were evaluated at day 7 and day 28 during the curing process. The addition of PSAs to the mortars caused an increase in mechanical resistance such that the 0.06% content presented better performance at day 7 of curing (4.07% higher) and at day 28 of curing (8.06% higher) when compared with the reference mortar (without the addition of PSAs) in the same curing periods. This work demonstrates that the addition of PSAs contributes to the hydration of a cementitious material, improving the mechanical resistance of the studied mortars.

Keywords: hydrogel; mortar; mechanical resistance; rice husk ash

1. Introduction

In recent years, there has been a great advance in technologies aimed at civil construction [1,2]. In this context, concrete and mortar are among the cementitious materials used the most in the structural part of civil construction [1] due to their positive characteristics, such as resistance, durability, moulding, accessibility, and the abundance of raw materials [2,3].

The durability of cementitious composites depends on several factors, such as the conditions that the material will be exposed to during its useful life (e.g., environmental action), and on factors related to its formulation, components, and structures, such as the type of cement, water/cement ratio [4,5], and additives to minimize crack formation [5]. The occurrence of cracks in a structure at an early age caused by shrinkage problems can also compromise the strength and durability of cementitious materials [5]. Shrinkage in cementitious materials occurs due to excessive water loss and a decrease in its resistance [6].

Among several methods which have been widely used to minimize the effect of shrinkage and simultaneously guarantee the resistance of cementitious materials, the application of so-called curing agents [6,7] can be highlighted. This technology is based on a basic procedure in which adequate humidity and temperatures are provided to the cementitious material to promote hydration of the material for a certain period [8]. Proper curing is crucial to achieving strength and therefore greater durability, especially for cementitious materials exposed to extreme environmental conditions at an early age [9]. This method can be divided into two groups: external curing and internal curing.

Internal curing occurs with a supply of water to the existing aggregates inside the cementitious material [7,10]. For this, water penetrates the cementitious material through its capillaries. Thus, the supply of water minimizes and compensates for water evaporation and increases the degree of hydration [11]. The internal curing of the cementitious compound can also occur through the addition of a curing agent in the formulation which will serve as an internal water reservoir. Such an agent can continuously and gradually release water as hydration occurs [7,12].

Several types of materials can be used as internal curing agents, among them being superabsorbent polymers (PSAs), rice husk ash (RHA), pre-moistened lightweight aggregates (LWAs), and fly ash (FA) [13,14]. Due to their characteristics, PSAs have become a very promising type of material for use as an internal curing agent, mainly due to their ability to absorb a significant amount of liquid, if immersed in aqueous environments and retain the liquid within its structure without dissolving [15]. The liquid absorbed by the PSAs can be released continuously and gradually over time to the outside (i.e., to the environment constituted by the elements of the cementitious material) [5]. It has been demonstrated that the use of PSAs can contribute to the hydration reaction of the material and minimize the shrinkage process through more adequate internal curing [16,17].

The use of an SAP as an internal curing agent is a subject which has been etensively studied due to its water release and desorption kinetics, which are directly related to the physical and chemical properties of the PSA [18,19]. In the literature, there are studies showing that the addition of an SAP improves the hydration of the medium, being able to prevent self-desiccation and mitigate autogenous shrinkage [19,20]. However, there is no consensus on how an SAP influences the mechanical property, as some studies showed that the addition of SAPs causes a reduction in mechanical strength [21,22], while others reported an increase in mechanical strength [23,24].

One hypothesis for the reduction in mechanical strength may be related to the particle size of the PSA incorporated into the cementitious medium [20,25], because the larger the SAP particle, the greater the amount of water absorbed or released, which can improve the hydration of the medium but can also lead to the formation of large pores, which can negatively affect the resistance of the material [19].

Some PSAs can absorb amounts of water up to 5000 times their own weight, in addition to having a three-dimensional network structure [15]. They are characterized by having a strong affinity for water due to the presence of hydrophilic groups, such as -OH, -COOH, $-CONH_2$, and $-SO_3H$, in their structure [17].

As for their classification, PSAs (hydrogels) can be classified as natural or synthetic, or they can be obtained from a combination of natural and synthetic polymers. PSAs have a high degree of flexibility due to their high water content [26–28]. Currently, the incorporation of natural polymers in the preparation of hydrogels for application in the areas of health and agriculture, for example, has been widely used to improve applicability, biocompatibility, and biodegradability [29]. Pectin hydrogels are often used [30,31] due to pectin being from natural sources. Pectin (Pec) is a polysaccharide of plant origin consisting of poly α 1-4-galacturonic acids found in the cell walls of terrestrial plants. It has a linear structure, is non-toxic and biocompatible, and has gelling, mucoadhesive, thickening, and emulsifying properties [32–36]. Due to these properties, this polysaccharide is widely used for the development of hydrogels [34]. Acrylamide is a vinylamide (CH_2=CH-$CONH_2$)—a white solid soluble in water—and is widely used for the synthesis of polyacrylamide through a polymerization reaction [37,38]. Rice husk (RH) is a byproduct obtained through the processing of rice. Its chemical composition consists of an organic part (74%) which includes cellulose, hemicellulose, and lignin and an inorganic part (26%) composed mainly of silicon dioxide (SiO_2) and other constituents, such as silicon oxide (SiO_2), alumina (Al_2O_3) sulphur trioxide (SO_3), iron oxide (Fe_2O_3), calcium oxide (Ca_2O), magnesium oxide (MgO), sodium oxide (Na_2O), and potassium oxide (K_2O). RH is considered a biomass, and therefore it is widely used for energy generation through its burning, which leads to the production of a residue called rice husk ash (RHA]) [39,40].

Considering the addition of a PSA as an internal curing agent, another advantage, aside from the release of additional water to make up for water lost through evaporation, is that it is also possible to control the rheological properties [14]. However, after the release of water by a PSA, formation of empty micropores in the hardened material may occur, thus increasing the porosity of the material and reducing its mechanical properties [14,30]. One way to minimize this problem is to incorporate materials such as rice husk ash (RHA), which after the release of water by a PSA can partially fill the pores formed and thus improve the mechanical resistance of the material.

RHA is a material that has a high silica (SiO_2) content, which makes it a pozzolanic material that, when added to the cementitious medium, can contribute to increasing the mechanical resistance. In addition, RHA is an agricultural byproduct, and seeking new applications for this byproduct contributes to reducing environmental problems caused by the incorrect disposal of this material [41–43].

The hypothesis evaluated in this work is that the incorporation of RHA would increase the mechanical properties in the hydrogel of PAAm, pectin, and RHA. In this work, hydrogels based on polyacrylamide (PAAm), pectin (Pec), and rice husk ash (RHA) are produced, in which the amounts of these components vary in different formulations through a 2^3 full factorial design in triplicate at the central point. The hydrogels are incorporated into formulations of cementitious materials (mortar) and characterized through measurements of the mechanical properties in periods of 7 and 28 days after preparation.

2. Materials and Methods

2.1. Materials

The hydrogels were prepared using apple pectin (CAS 900-69-5, Sigma-Aldrich, Saint Louis, MO, USA) with a degree of methoxylation of 50–75%; acrylamide (Fisher Scientific, USA); potassium persulfate (Synth, Diadema, Brazil); N,N′-methylenebisacrylamide 99.5% (CAS 110-26-9, Fisher Scientific, Waltham, MA, USA); and rice husk, which was purchased from the municipality of Boqueirão do Piauí, PI, Brazil. Portland cement CP II-F and sand (Brazilian normal in 4 fractions: coarse (maximum size: 1.19 mm); medium coarse (maximum size: 0.59 mm); medium fine (maximum size: 0.297 mm), and fine (maximum size: 0.149 mm)) were used to produce the mortar according to ABNT/NBR 7214/2015 [44].

2.2. Methods

2.2.1. Preparation of RHA and Synthesis of the Hydrogels

The rice husk ash (RHA) was obtained through calcination in a muffle furnace at 600 °C for 2 h and cooled to room temperature inside the muffle for 24 h. After this procedure, the ash produced was ground and sieved through a 325 mesh (44 μm) sieve [45].

A 2^3 full factorial design in triplicate at the central point was used to produce hydrogels. The Q value was used to analyze the influence of three factors: the acrylamide (AAm) content, Pec content, and RHA content. The levels of each factor were adjusted based on preliminary studies, and the values for the upper (+), middle (0), and lower (−) levels are shown in Table 1. The levels of the three factors were combined, resulting in 11 experiments, as described in Table 2, with experiments 9a, 9b, and 9c being replicates of the central point.

For the hydrolysis of the hydrogels, 40 mL of 0.5 mol L^{-1} NaOH solution was used per 1 g of dry hydrogel. In a hood, the solution was magnetically stirred for 1 h at 50 °C. Then, the hydrogels were washed with distilled water until reaching a pH of 7 and subsequently dried in an oven at 60 °C for 24 h [41]. In the preparation of hydrogels in each of the different formulations (different experimental conditions), the levels of the factors described in Table 1 and their combinations, as described in Table 2, were used.

For the synthesis of hydrogels, the pectin was first dissolved in distilled water (necessary volume to maintain the total concentration at 136 g L^{-1}) at 50 °C. The solution was then kept at room temperature and stirred for 24 h. After pectin solubilization, rice husk ash was added to obtain a suspension. For this, the system was kept under agitation (by means of a magnetic stirrer) at 70 °C. Then, the required amount of acrylamide (AAm), 0.04 g of N,N′-methylenebisacrylamide (MBAAm), and 0.08 g of sodium persulfate were added, with an interval of 10 min between the addition of each reagent in the sequence.

Table 1. Factors and respective levels used in factorial planning.

Factors (g)	Lower Level (−)	Central Point (0)	Upper Level (+)
AAm	1.5	2.5	3.5
Pec	1.0	2.0	3.0
RHA	0.1	0.2	0.3

Table 2. Factorial design matrix used for the synthesis of hydrogels.

Run	AAm Content	Pec Content	RHA Content	H_2O (mL)
H1	−1	−1	−1	19.12
H2	+1	−1	−1	33.82
H3	−1	+1	−1	33.82
H4	+1	+1	−1	48.3
H5	−1	−1	+1	20.59
H6	+1	−1	+1	35.29
H7	−1	+1	+1	35.29
H8	+1	+1	+1	50.00
H9a	0	0	0	34.55
H9b	0	0	0	34.55
H9c	0	0	0	34.55

The hydrogels obtained were divided into small pieces and left immersed in distilled water to be washed (removal of reagents not incorporated into the hydrogel). The water was renewed every 1 h for a total period of 6 h. The obtained material was then dried in an oven at 60 °C for 24 h [31]. In addition, the hydrogels which were applied to the cementitious compound, after drying, were crushed and sieved until obtaining the granulometry of a 250 mesh.

2.2.2. Preparation of Mortars and Cement Pastes

Four formulations of mortars and cement pastes were prepared with a water/cement ratio of 0.48 according to NBR 7215 (ABNT/1996) [46] and traces at a ratio of 1:3 (cement: sand). The hydrogel used was H4 (formulation 4; see Table 2), with 0.03%, 0.06%, and 0.1% hydrogel in relation to the mass of the cement. A reference formulation (Ref) without the addition of hydrogel was also implemented following the same procedure, and it was used for comparison in the characterization tests. The as-prepared hydrogels were left to swell for 20 h in tap water before being used to prepare the cementitious material. The fine aggregate was normal sand inserted at 4 different granulometries: fine sand, medium-fine sand, medium-coarse sand, and coarse sand.

Cement pastes were prepared by mixing water and cement with and without the addition of hydrogel. The pastes containing 0.03%, 0.06%, and 0.1% hydrogel were labeled P-0.03, P-0.06, and P-0.1, respectively, and the one without hydrogel was labeled P-Ref.

To prepare the mortar, the mixer was first turned on, and fine aggregates and cement were added to the tank. The mixture was maintained at a low speed for 60 s, and then water was added with the hydrogel swollen until equilibrium. Then, the material was maintained at a high speed for 60 s. After this time, the mixer was turned off, and the material which remained on the sides of the vat was removed with a spatula and placed inside the vat. Then, the materials were mixed for another 2 min at high speed.

After preparing the mortar, the test specimens were moulded into a cylindrical shape 50 mm in diameter and 100 mm in height. They were prepared and labeled as Mort0.03, Mort0.06, Mort0.01, and Mort-Ref. After a period of 24 h, the samples were cured at room temperature (25 °C) and with a relative humidity of 72%.

2.3. *Characterization*

2.3.1. Fourier Transform Infrared Spectroscopy (FTIR)

The FTIR spectra of the hydrogels, RHA, and cement paste were obtained in a Bruker Vertex 70 spectrometer in the range of 400–4000 cm^{-1}, with a resolution of 4 cm^{-1}. Powdered samples were prepared using the pelleting technique in 1% KBr (64 scans).

2.3.2. X-Ray Diffraction (XRD)

Samples of rice husk ash, cement paste, and hydrogels were characterized by X-ray diffraction in a Shimadzu XRD-600 (Kyoto, Japan) using a CuKα radiation source, with a voltage of 40 kV and current of 30 mA. Data were collected at a speed of 1° min^{-1}, whose sweep range was (2θ) from 3 to 75°.

2.3.3. Scanning Electron Microscopy (SEM)

The morphologies of the hydrogels were analyzed using a scanning electron microscope (FEI, model Quanta 250, Pittsburgh, PA, USA). To obtain the images, the samples were covered with a thin layer of gold of approximately 50 nm. The images were obtained with an accelerating voltage of 20 kV. The average pore diameters of the hydrogels were analyzed using Image-J® 1.8.0 software by manually delimiting the pore contours observed in the SEM images.

2.3.4. Thermogravimetric (TG and DTG)

The samples of the cement pastes were characterized in a Shimadzu DTG-60 to analyze the hydration process of the pastes. The tests were carried out at room temperature initially and up to a final temperature of 600 °C, with a heating rate of 10 °C and a N_2 flow at 100 mL min^{-1}.

2.3.5. Water Swelling, Desorption Test, and Testing of the Drying and Swelling Effects of Hydrogels

Swelling tests were used to study the water absorption of the hydrogels immersed in different media, including distilled water, a buffer solution (pH: 4, 7, and 9), and a saline solution (NaCl). All solutions (buffer and saline) had a constant ionic strength of 0.1 mol L^{-1}. In this way, dry hydrogels with a mass of approximately 0.06–0.1 g were immersed in a beaker with 30 mL of a solution or distilled water. The hydrogels were removed at different time periods (every 1 h for a total period of 8 h at 24 h, 48 h, and 72 h). In each case, excess water from the sample was removed, and then the hydrogel was weighed on an analytical balance. For each hydrogel formulation (Table 2), three trials were performed (n = 3). The swelling capacity (measured with the Q parameter) could be calculated according to Equation (1), where m_1 is the mass of the dry material and m_2 is the mass of the swollen hydrogel:

$$Q = [(m_2 - m_1)/m_1] \quad (1)$$

For the desorption assay, the hydrogels swollen to equilibrium (m_2) were placed in an oven at 30 °C. The samples were weighed at different drying time intervals. For each hydrogel formulation (Table 2), tests were performed in triplicate (n = 3).

The process of the repeat effect of drying and swelling of the hydrogels was verified in an aqueous medium. The dried hydrogel was immersed in water, and the absorption was determined as previously described above. The material was then dried at 60 °C for 24 h and swollen again. The process was then repeated 5 times.

2.3.6. Resistance Test of Cementitious Materials

The compressive strength of the mortar was measured following the recommendations of NBR 7215 (ABNT/1996) [46] at ages of 7 and 28 days, using an EMIC Pc 100 testing machine (mechanical press). For the compression test, the test specimen was placed in the center of the lower plate of the mechanical press, and then the equipment was turned on, exerting force until the test specimen ruptured.

For each formulation, 4 mortar test specimens were used, testing each prepared mix ($n = 4$). The compressive strength result was obtained by averaging the individual resistance values of each test specimen of the same age through the quotient between the force recorded by the equipment and the area of the cross-section of the test specimen where the force was applied.

3. Results

3.1. Kinetic Behavior of Water Absorption by Hydrogels

Figure 1 shows the degree of swelling (Q) of the hydrogels in distilled water for different immersion times. For all samples, the degree of swelling (Q) increased rapidly in the first few hours. It can be observed that both the H3 and H5 hydrogel formulations reached the equilibrium value after 8 h of immersion, while the other hydrogel formulations reached approximately 90% of the equilibrium in 24 h. Thus, the H3 and H5 formulations showed quicker swelling processes and reached equilibrium faster. In addition, the other hydrogel formulations showed a slower capacity to absorb water at equilibrium. The swelling degrees of H4 and H8 (Q, in g/g) were 55.72 and 47.77, respectively, which were higher compared with the other hydrogels. Therefore, it is inferred that the Q values and the time required to reach swelling equilibrium depend on the formulation of each hydrogel [47]. Thus, it can be said that the H4 hydrogel presented better water absorption due to the higher starch and pectin contents in its composition, which means that the number of hydrophilic groups was greater, thus favoring water absorption.

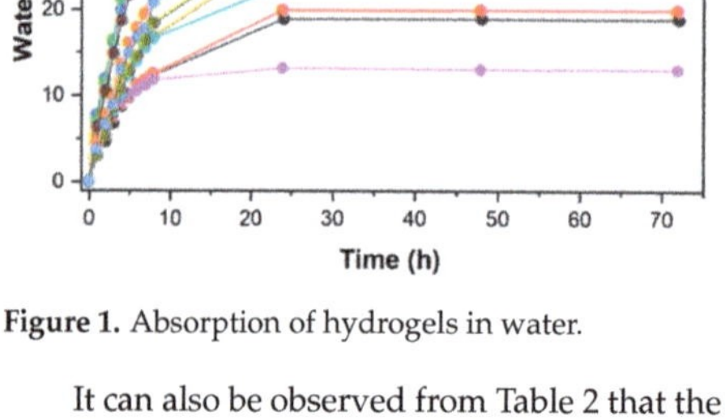

Figure 1. Absorption of hydrogels in water.

It can also be observed from Table 2 that the H4 and H8 hydrogels had similar formulations, differing only in the amount of RHA, which emphasizes that the H8 formulation hydrogel had a higher amount of RHA. This explains the reduced swelling of the hydrogel of the H8 formulation compared with that of the H4 formulation. Such a reduction in the Q value may be related to the presence of a greater amount of RHA which filled the hydrogel pores and, at the same time, reduced the elasticity of the matrix [45], aspects which led to a decrease in the degree of swelling.

Since the H4, H8, and H9 hydrogels showed higher degrees of swelling if compared with the other hydrogels prepared in this work, these were chosen to carry out the subsequent analyses with the objective of evaluating which was the best hydrogel to be applied in cementitious materials.

3.2. Kinetic Behavior of Water Absorption of Hydrogels at Different pH Levels and in Saline Solution

Figure 2a shows the dependencies of the degree of swelling of hydrogels H4, H8, and H9 at different pH values, and the dependence of the degree of swelling on the saline solution is shown in Figure 2b. It can be seen in Figure 2a that at a pH of four, the H9 hydrogel had a lower Q value than the H4 and H8 hydrogel formulations. Furthermore, at a pH of four, the three hydrogels showed lower water absorption values, and as the pH of the solution increased from 4 to 7 and from 7 to 9, subsequent increases in the Q value occurred, which may be explained as follows. In an acid medium (pH of four or less), the carboxyl groups are mostly in protonated form (-COOH), and thus there is little (or no) electrostatic repulsion between them [48,49]. It is emphasized that the pKa of carboxylic groups is close to 4.0–4.5 [31,50]. Thus, at pH levels of seven and nine, the carboxylic groups were in dissociated form ($-COO^-$). Thus, at pHs of seven and nine (pH > pKa), the carboxylate anions presented electrostatic repulsion among themselves. This induced a greater inflow of water (or aqueous liquids) into the hydrogel matrix to increase the distance between the negatively charged groups and thus reduce electrostatic repulsion, leading to a greater degree of swelling. Thus, the hydrogel matrix consequently swelled more and expanded [31,45].

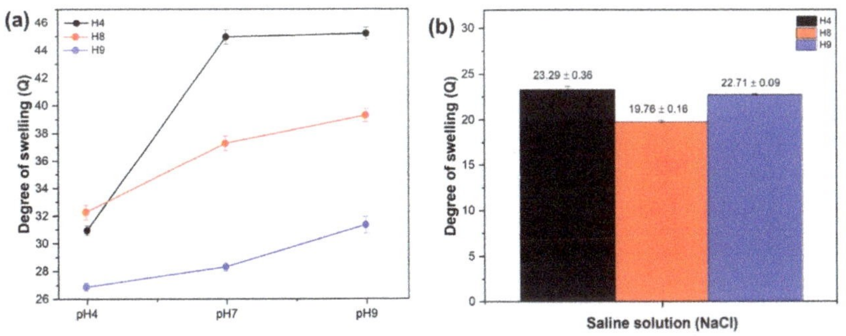

Figure 2. (**a**) Degree of swelling of hydrogels at different pH levels and (**b**) in saline solution.

The influence of the salinity of the solution on the degree of swelling of the hydrogels (in 0.1 mol L^{-1} of NaCl solution) is shown in Figure 2b. Tests in a saline environment were performed for evaluating the influence of salts on the swelling degree of the hydrogels through the factor f in Equation (2). A value close to 1.0 for f indicates a weaker effect of salts on the swelling properties of the hydrogel. Therefore, the H4 hydrogel was the best while H9 was the worst when the hydrogels were swollen in 1 mol L^{-1} aqueous NaCl. The swelling capacity of the hydrogels in the saline solution decreased in relation to the swelling in distilled water (Figure 1) and the buffer (Figure 2b), and the degree of swelling of samples H4, H8, and H9 in saline at its retention equilibrium was about 23.296 ± 0.36, 19.767 ± 0.16, and 22.713 ± 0.09, respectively. It can be explained that this reduction was due to the sodium cations (Na^+) interacting in the saline solution, with the pectin carboxylate groups forming ionic complexes, which provided a decrease in the electrostatic repulsion in the polymeric matrix [51]. With the decrease in electrostatic repulsion, less water would be required to minimize repulsion, and therefore less water needed to diffuse into the hydrogel, and the degree of swelling would be lower. In addition, the sensitivity factor (f) of the hydrogels to salt is defined by Equation (2) [52,53]:

$$f = 1 - [\text{Qsal}/\text{Qwater}] \quad (2)$$

where Qsaline is the degree of swelling in the saline solution and Qwater is that in the distilled water. The f values of the hydrogels (Table 3) show that the H9 hydrogel had a lower saline effect compared with the H4 and H8 hydrogels, which had similar saline effects. This is an indication that the ash content did not affect the value of f.

Table 3. The ionic strength function for H4, H8, and H9 hydrogels.

Hydrogel	H_2O (g)	NaCl (1 mol L^{-1}) (g)	f
H4	55.723	23.296	0.581
H8	47.777	19.767	0.587
H9	36.535	22.713	0.371

3.3. Desorption of Swelling Water from Hydrogels

The release kinetics of the water adsorbed by hydrogels H4, H8 and H9, are shown in Figure 3. Each point in this figure represents an average obtained from the triplicates ($n = 3$). The water release from the H4 and H8 hydrogel formulations was approximately equal over time (up to 48 h). Therefore, the points almost completely overlapped. It can be observed that about 50% of the water adsorbed by the H4 and H8 hydrogels was released in the first 7 h of the experiment, while the H9 hydrogel in the same period released approximately 35% of the total swelling water contained therein. Furthermore, about 95% of all of the water contained in the hydrogel matrices was released within 24 h. In 48 h, the three hydrogels almost completely lost the adsorbed water, but it is known that it is quite difficult for the hydrogel to lose 100% of the water contained in its matrix [17]. It can be observed that the H4 and H8 hydrogels presented rather similar water release profiles throughout the experiment time (0–48 h). The formulations to obtain these hydrogels differed only in their RHA content, with the H4 hydrogel having an RHA content factor at a lower level and that of H8 being at an upper level (Table 2).

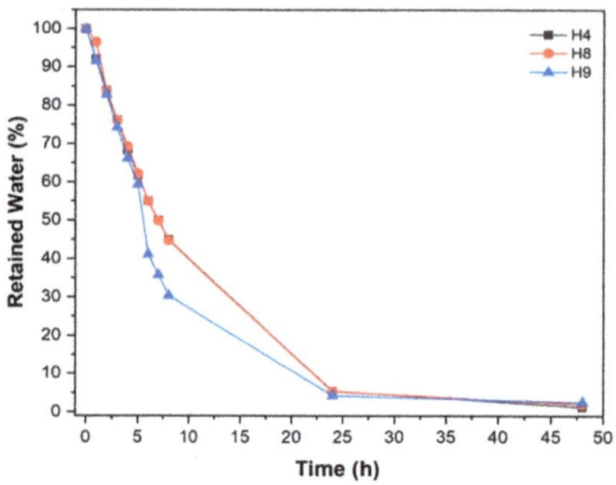

Figure 3. Release profiles of adsorbed water in hydrogels H4, H8, and H9.

3.4. Testing of the Drying and Swelling Effects of Hydrogels

The water's release profiles were measured for evaluating the time in which the water molecules diffused out of the hydrogel when it was maintained in an oven at a controlled temperature. When inside the cementitious material, the time for the water to diffuse out of hydrogels would be higher. Therefore, it was expected that this could be related to the time in which the water acted as a curing agent inside the cementitious material. In the H4 and H8 hydrogel formulations, the AAm and Pec content factors were maintained at their upper levels. Thus, it appears that the RHA content did not influence the kinetics of water release. Knowledge of the water release profile by hydrogels is a highly important aspect for obtaining an adequate application of the hydrogel as a curing agent in cementitious materials. As was already mentioned, the continuously and gradually released water would contribute to cement hydration and thus improve the internal curing process.

Figure 4 shows the effect of repeating the swelling and drying cycle of the hydrogels on the degree of swelling. It was observed that all samples showed an initial decrease in the swelling degree value (Q) and only stabilized after the fourth drying and swelling cycle. When the hydrogel met water molecules, it swelled and could expand to many times its initial volume. Thus, repetition of the drying and swelling cycle affected the water absorption capacity [45]. The H4 hydrogel showed greater reduction (59%) in the Q value after the fifth swelling and drying cycle, while the H8 and H9 hydrogels showed a reduction of 53% and 56%, respectively.

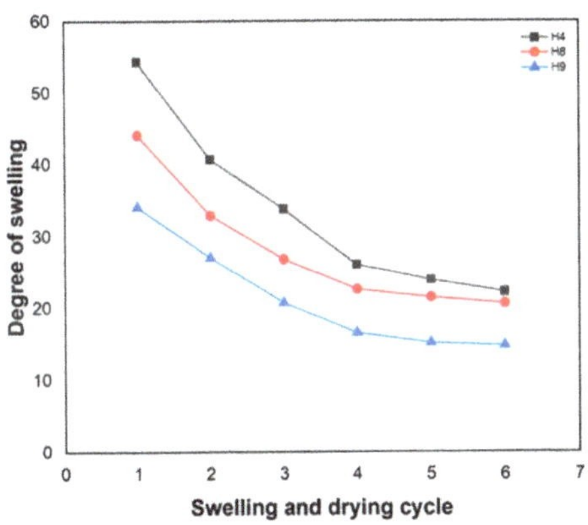

Figure 4. Effect of swelling and drying cycles of hydrogels on water absorption capacity.

3.5. FTIR, XRD, and Morphology Analysis of Hydrogels

The FTIR spectra of AAm, RHA, and Pec are shown in Figure 5a. In the FTIR spectrum of the RHA bands at 1091, 786 and 466 cm^{-1}, symmetric and asymmetric elongations and angular deformations of the Si-O-Si bonds of the silica present in the material were observed [45]. The bands at 3492 and 1614 cm^{-1} can be attributed to the O-H stretching and deformation vibrations of the silanol group, respectively [45–54]. The AAm FTIR spectrum showed characteristic bands at 3377 and 3195 cm^{-1}, which can be attributed to asymmetric and symmetric elongation of the amide (-NH) groups. The bands at 1674 and 1604 cm^{-1} corresponded to C=O elongation and the N-H bond [55]. In the pectin spectrum, bands at 3464 and 2951 cm^{-1} were observed, which refer to the stretching vibration of -OH and -CHx, respectively.

The bands at 1751 cm^{-1} and 1635 cm^{-1} can be attributed to the elongation vibrations of the C=O groups present in the -COOCH group. In addition, the band at 1022 cm^{-1} was due to –CH–O–CH– elongation [33,56,57]. In addition to the bands already mentioned for PAAm, RHA, and Pec, it was observed that the spectra of hydrogels H4, H8, and H9 (Figure 5b) were similar, showing bands at 3456 cm^{-1} and 3207 cm^{-1} which corresponded to the -OH and -NH stretching vibrations. The band at 2937 cm^{-1} refers to the C-H stretching vibration, and the band at 1666 cm^{-1} refers to the C=O stretching vibrations of the amide group of PAAm [58]. There was also the presence of a band at 466 cm^{-1}, attributed to the Si-O-Si binding of RHA. Therefore, it can be concluded that the appearance of these bands showed that the hydrogels containing PAAm, pectin, and rice husk ash (RHA) were successfully synthesized.

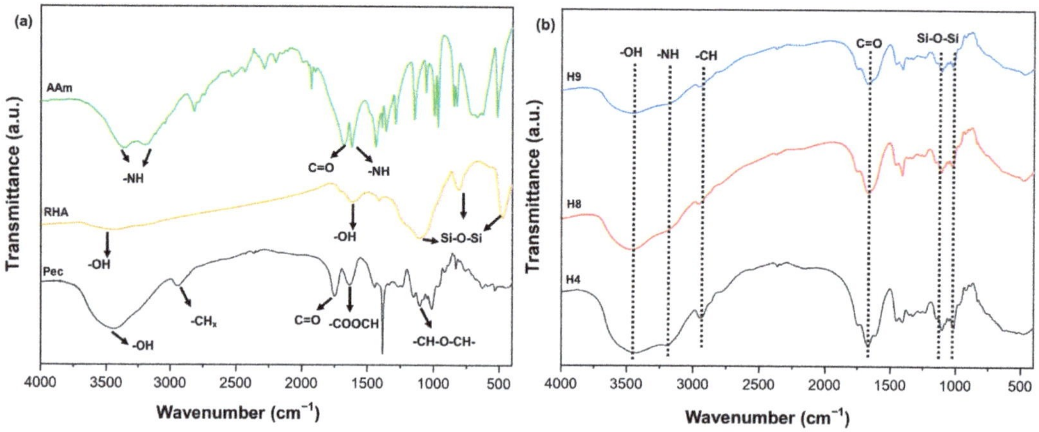

Figure 5. FTIR spectra of (**a**) Pec, RHA, and Aam and (**b**) H4, H8, and H9 hydrogel formulations.

Figure 6 presents XRD diffractograms of pristine rice husk ash and hydrogels H4, H8, and H9. The diffraction pattern shows that the ash diffraction pattern had no defined diffraction peaks; only a broad peak at $2\theta = 22.3°$ is displayed, showing the non-crystalline phase of silica [29,45]. The diffractogram of the hydrogels showed an amorphous structure which is characteristic of a disorganized system and the presence of a peak referring to the amorphous silica of the RHA. Thus, it can be concluded that the RHA was incorporated into the polymer matrix of the hydrogel during the gelation process.

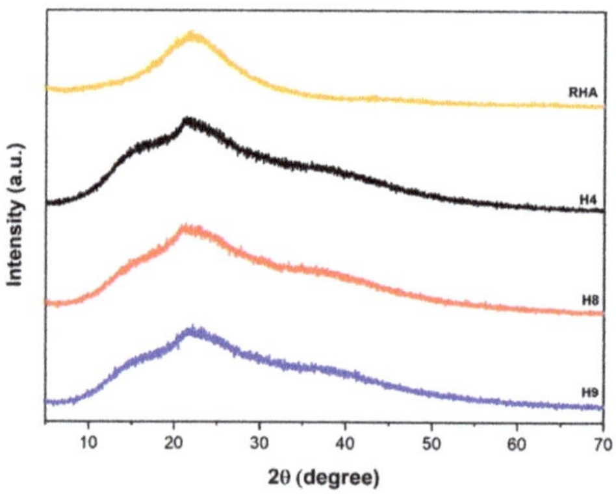

Figure 6. XRD diffractograms of RHA and H4, H8, and H9 hydrogel formulations.

SEM images of the H4, H8, and H9 hydrogels are shown in Figure 7a–c. All hydrogels exhibited porous structures with irregular shapes, in addition to pores of different sizes which were distributed throughout the hydrogel matrices. The distributions of the average size for 19 pores ($n = 19$) of the hydrogels, which were calculated through their areas, are shown in Figure 7d–f. The average pore sizes of the H4, H8, and H9 hydrogels were 20.82, 30.13, and 29.75 μm, respectively. It was observed that H8 showed a larger average pore size in relation to the other two hydrogel formulations. This indicates that the introduction of RHA into preparation of the hydrogels increased the size of the pores, making a structure with pores larger than those of the others.

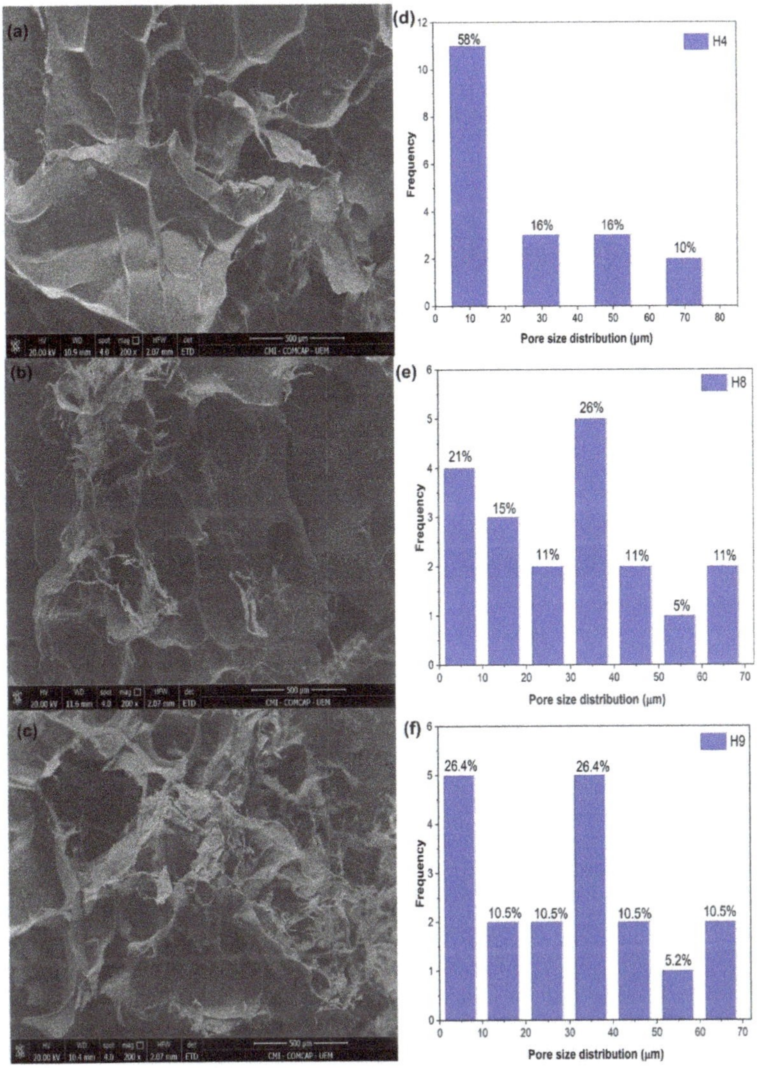

Figure 7. SEM micrographs of the hydrogels: (**a**) H4, (**b**) H8, and (**c**) H9. Magnification: 200×. (**d–f**) Pore size distribution.

3.6. FTIR, XRD, and TG/DTG Analysis of Cement Pastes

Figure 8 presents the FTIR spectra of the cement pastes after 7 days of curing. The cement pastes containing hydrogels presented FTIR spectra quite similar to the reference paste. The FTIR spectra of the cement pastes had bands at 3460 and 1672 cm^{-1}, which can be attributed to symmetric and asymmetric elongation related to vibration and bending of the O-H groups of water molecules [59,60]. The band observed at 3649 cm^{-1} is characteristic of calcium hydroxide ($Ca(OH)_2$) due to the stretching vibration of the O-H bond [59,61]. The bands at 1456 and 873 cm^{-1} are related to elongation vibrations of the C-O bonds in calcium carbonate [62,63]. The spectra of the pastes had bands at 979 and 711 cm^{-1}, which can be attributed to bending vibrations of the Si-OH bonds in C_2S and C_3S [59,62,63]. According to the analyses carried out in the FTIR spectra, it was possible to identify the hydrated compounds present in the cement samples. Furthermore, the modified cement pastes containing different hydrogel contents contained the same hydrated compounds as the reference paste.

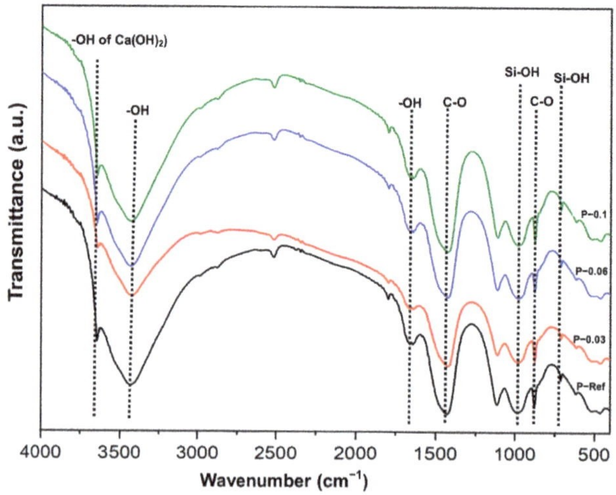

Figure 8. FTIR of cement pastes at day 7.

The hydrogel chosen to be applied to the cementitious compounds was H4, based on the factorial design and because it presented better results in the swelling tests and exhibited a smaller pore size. After water desorption, the PSAs left empty pores in the hardened cementitious material. Thus, the smaller the pore size of the PSAs, the smaller the voids left by them. In Figure 9, the XRD profiles of the cement pastes at days 7 and 28 of curing are presented. The hydrated compounds identified in all cement pastes were calcium hydroxide ($Ca(OH)_2$), calcite ($CaCO_3$), ettringite, hydrated calcium silicate (C-S-H), belite (C_2S), and alite (C_3S). The diffractograms obtained for the reference pastes (without the addition of hydrogels) were like those of the pastes containing the hydrogels over time (at days 7 and 28 of curing), and this result shows that the use of different hydrogel contents in the formulation of the mortar did not alter the crystallographic structure of the hydrated cementitious material. However, it can be observed that the $Ca(OH)_2$ peak increased with the addition of different hydrogel contents in the paste.

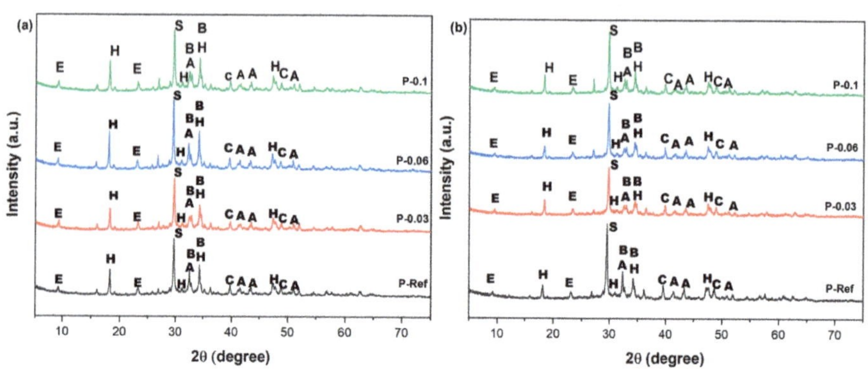

Figure 9. XRD of cement pastes at 7 days and 28 days. (**a**) Paste aged 7 days. (**b**) Folder aged 28 days. Note: calcium hydroxide (H), ettringite (E), hydrated calcium silicate (S), belite (B), alite (A), and calcite (C).

Considering the XRD profiles of the pastes which were analyzed at days 7 and 28 of curing, both were structurally quite similar. However, for the intensities of the peak of calcium hydroxide in the region of 2θ equal to 18.2 and 34.0°, a reduction occurred in the period of 28 days of curing when compared with the intensity of the same peak in the XRD profile of the paste at day 7 of curing. This was due to the pozzolanic reaction consuming $Ca(OH)_2$ during the cement hydration process due to the presence of silica [12,64,65]. Through the diffractograms of the pastes, the main peaks which were found for calcium hydroxide at 2θ = 18.2, 31.8, 34.2, and 47.3°, hydrated calcium silicate at 2θ = 29.6°, ettringite at 2θ = 9.4 and 23.1°, belite at 2θ = 32.3 and 34.2°, alite at 2θ = 32.3, 41.1, 43.3, and 51.1°, and calcite at 2θ = 39.6 and 48.9° were responsible for the durability of the cementitious compound [64,66–68].

Figure 10 shows the TG and DTG curves of the cement pastes containing hydrogels and the references in which they were analyzed after 90 days of curing. It can be observed that all of the pastes presented mass loss at temperatures of 25–105 °C, which corresponds to evaporation of the absorbed water [69,70]. Furthermore, it was observed that the TG curves of the hydrogel-containing pastes were like the reference paste, indicating that no new events were found. The DTG curves of the pastes showed peaks at 78 °C associated with the release of water in the pores and the dehydration of the hydrated compounds, namely hydrated calcium silicate and ettringite [69,71,72], having a peak with a maximum at 430 °C corresponding to the dehydration of the $Ca(OH)_2$ [69,70].

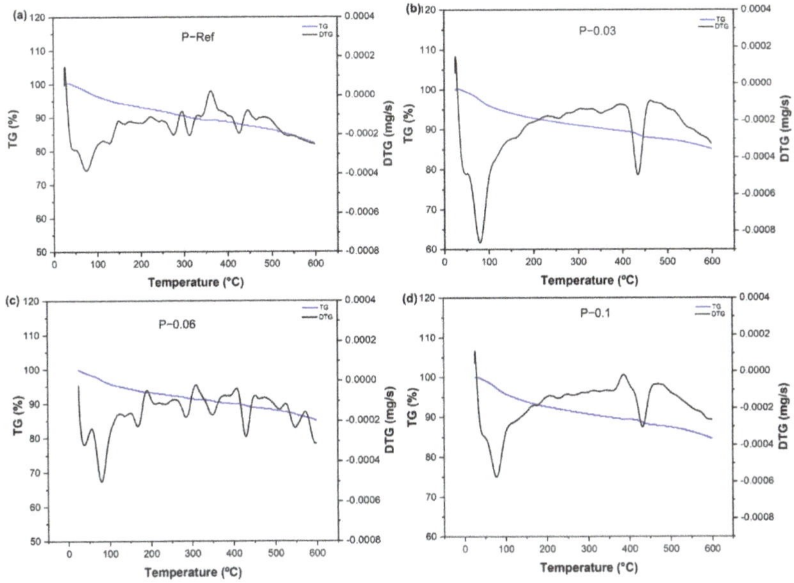

Figure 10. TG and DTG curves of cement pastes: (**a**) P−Ref, (**b**) P−0.03, (**c**) P−0.06, and (**d**) P−0.1.

3.7. Compressive Strength of Cement Mortar

The results for the compressive strength of cement mortars with the addition of different hydrogel contents (0.03%, 0.06%, and 0.1%) are shown in Figure 11. It can be seen that in the 7 day curing period, sample Mort006 showed the greatest increase in mechanical strength (an increase of 4.07%) and Mort01 showed a slight increase (an increase of 0.47%) in compressive strength compared with the reference mortar (Mort-Ref) in the same period. However, in sample Mort003, there was a 5.3% decrease in strength compared with the reference mortar.

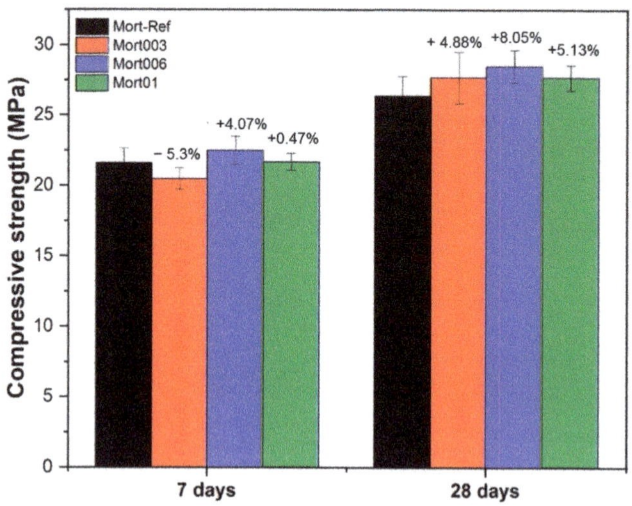

Figure 11. Compressive strength of cement mortars containing different hydrogel contents for aging at days 7 and 28.

It can be observed in Figure 11 that in the 28 day curing period, the compressive strengths of Mort003, Mort006, and Mort 01 increased by 4.88%, 8.05%, and 5.13%, respectively, in relation to the mechanical strength of Mort-Ref. In addition, it can be noted that in the Mort003 sample, there was a significant increase with 28 days of curing in relation to the resistance at 7 days. This may occur because the hydrogel releases water in a controlled manner, favoring the curing process of the mortar [31]. In addition, the reduction in strength may also be related to the voids in the mortar structure created by the hydrogels [20].

Through the results obtained in the resistance tests, it was possible to conclude that the addition of different hydrogel contents in the mortar did affect its resistance, since all of the samples showed an increase at day 28 of curing when compared with the reference one. In addition, the Mort006 sample was the one that showed the greatest increase in resistance over time in relation to the other grades studied.

To better analyze whether the hydrogel content and the age of the samples interfered with the compressive strength results, analysis of variance (ANOVA) was carried out at a confidence level of 95%, and the results are presented in Table 4, where it should be noted that the age and hydrogel content were statistically significant.

Table 4. Analysis of variance of the compressive strength test data of the mortars, considering the main factors of age and the hydrolyzed hydrogel content.

Effect	SQ *	DF **	MQ ***	F	p Value	Result
Age	1.99×10^{14}	11	1.99×10^{14}	141.5	1.27×10^{-7}	Significant
Hydrogel	2.00×10^{14}	6	4.00×10^{13}	18.2	5.30×10^{-6}	Significant
Residue	1.55×10^{13}	-	1.41×10^{12}	-	-	-

* SQ = quadratic sum. ** DF = degrees of freedom. *** MQ = mean of squares.

Therefore, based on the results obtained in this work, it can be observed that the incorporation of hydrogels based on PAAm, Pec, and RHA in cementitious materials (mortar) significantly improved the compressive strength. This increase may be associated with a good retention capacity and the subsequent release of stored water at the right time for the hydration process, considering the reactions which occurred during the setting time [73,74], thus contributing to the hydration reaction of the cementitious material. As expected, the influence of the age of curing affected the compressive strength more than the hydrogel.

4. Conclusions

Hydrogels based on polyacrylamide and pectin and containing rice husk ash were successfully synthesized, as can be seen from the FTIR, XRD, and SEM images. The hydrogel obtained from the H4 formulation showed a better Q value. In addition, the H4 and H8 samples showed a controlled and quite similar release of water over time. This makes these hydrogels rather interesting materials to apply to cementitious material as an internal curing agent, as it releases water at the right time for the setting process and thus contributes to cement hydration and improving its mechanical strength.

In this work, the influence of the addition of different contents of hydrogel formulation H4 for application in cement mortar at days 7 and 28 of curing was also studied. The result showed that the addition of hydrogels improved the mechanical resistance of the mortar, and the addition of 0.06% presented a better result for resistance to compression at day 7 of curing (4.07% higher) and at 28 days of curing (8.06% higher) when compared with the reference mortar (without the addition of PSAs). Furthermore, the FTIR, DRX, and TG analyses showed that the addition of the hydrogels in the pastes did not change the crystallographic structure or the chemical characteristics of the cement.

To the best of our knowledge, this is the first time that hydrogel formulations using pectin, rice husk ash (RHA), and crosslinked polyacrylamide were used as curing agents for cementitious materials. Beyond the technological use of hydrogel as a curing agent, this methodology is also interesting because it uses a natural product (pectin) and RHA as byproduct from rice agro-industries. Therefore, it should be also seen as a contribution to the environment. In this way, it can be concluded that the production of polyacrylamide hydrogels containing RHA and pectin can be used as an internal curing agent for cementitious media, in addition to adding commercial value to rice husk and preventing it from being discarded incorrectly and thus causing environmental problems.

Author Contributions: Data curation, investigation, methodology, software, validation, and writing—original draft, R.H.S.R.; data curation, investigation, methodology, software, validation, and writing—original draft, E.A.d.A.; investigation, methodology, software, and validation, F.R.K.; data curation, software, validation, co-supervision, and writing—original draft, E.C.S.-F.; conceptualization, funding acquisition, supervision, writing—original draft, and writing—review and editing, E.C.M. All authors have read and agreed to the published version of the manuscript.

Funding: This research was funded by the Conselho Nacional de Desenvolvimento Científico e Tecnológico (CNPq) (Grant #406973/2022-9, #406050/2022-8, #307429/2018-0, and #408767/2021-9).

Institutional Review Board Statement: Not applicable.

Informed Consent Statement: Not applicable.

Data Availability Statement: The original contributions presented in the study are included in the article, further inquiries can be directed to the corresponding author.

Conflicts of Interest: The authors declare that they have no known competing financial interests or personal relationships that could have appeared to influence the work reported in this paper.

References

1. Mechtcherin, V. Use of superabsorbent polymers (SAP) as concrete additive. *RILEM Tech. Lett.* **2016**, *1*, 81–87. [CrossRef]
2. De Brito, J.; Saikia, N. *Recycled Aggregate in Concrete: Use of Industrial, Construction and Demolition Waste*, 2013 ed.; Springer: London, UK, 2012; pp. 1–448.
3. Hazarika, A.; Hazarika, I.; Gogoi, M.; Bora, S.S.; Borah, R.R.; Goutam, P.J.; Saikia, N. Use of a plant based polymeric material as a low cost chemical admixture in cement mortar and concrete preparations. *J. Build. Eng.* **2018**, *15*, 194–202. [CrossRef]
4. Helene, P.; Andrade, T. Concreto de Cimento Portland. In *Materiais de Construção Civil e Princípios de Ciência e Engenharia de Materiais*, 3rd ed.; Isaia, G.C., Ed.; IBRACON: São Paulo, Brazil, 2017; Volume 2, pp. 905–944.

5. Wu, L.; Farzadnia, N.; Shi, C.; Zhang, Z.; Wang, H. Autogenous shrinkage of high performance concrete: A review. *Constr. Build. Mater.* **2017**, *149*, 62–75. [CrossRef]
6. Urgessa, G.; Choi, K.B.; Yeon, J.H. Internal Relative Humidity, Autogenous Shrinkage, and Strength of Cement Mortar Modified with Superabsorbent Polymers. *Polymers* **2018**, *10*, 1074. [CrossRef]
7. Mechtcherine, V.; Dudziak, L. Effects of Superabsorbent Polymers on Shrinkage of Concrete: Plastic, Autogenous, Drying. In *Application of Superabsorbent Polymers (SAP) in Concrete Construction State of the Art Report Prepared by Technical Committee 225-SAP*; Mechtcherine, V., Reinhardt, H.W., Eds.; Springer: New York, NY, USA, 2017; Volume 2, pp. 63–98.
8. Wang, K.; Cable, J.K.; Zhh, G. Investigation into Improved Pavement Curing Materials and Techniques: Part 1 (Phases I and II) (No. Iowa DOT Project TR-451). Iowa State University. Center for Transportation Research and Education. Available online: https://rosap.ntl.bts.gov/view/dot/24353 (accessed on 2 October 2024).
9. James, T.; Malachi, A.; Gadzama, E.W.; Anametemok, A. Effect of curing methods on the compressive strength of concrete. *Niger. J. Technol.* **2011**, *30*, 14–20.
10. Weiss, J.; Bentz, D.; Schindler, A.; Lura, P. Internal Curing, Structure Magazine. Available online: https://tsapps.nist.gov/publication/get_pdf.cfm?pub_id=910059 (accessed on 2 October 2024).
11. Bentz, D.P.; Jensen, O.M. Mitigation strategies for autogenous shrinkage cracking. *Cem. Concr. Compos.* **2004**, *26*, 677–685. [CrossRef]
12. Manzano, M.A.R. Experimental Study of High Strength Cementitious Materials Modified with Superabsorbent Polymers (SAP) as Internal Curing Agents. Ph.D. Thesis, Department of Civil and Environmental Engineering, University of Brasilia, Brasília, Brazil, 2016.
13. Manzano, M.A.R.; Silva, E.F.; Lopes, A.N.M.; Tolêdo Filho, R.D. Actuating mechanism of Superabsorbent Polymers as internal curing agents to mitigate autogenous shrinkage in High Strength Concrete (HSC)—State of the art. *Rev. Mater.* **2021**, *26*, 12956.
14. Danish, A.; Mosaberpanah, M.A.; Salim, M.U. Robust evaluation of superabsorbent polymers as an internal curing agent in cementitious composites. *J. Mater. Sci.* **2021**, *56*, 136–172. [CrossRef]
15. Jensen, O.M.; Hanse, P.F. Water-entrained cement-based materials: I. Principles and theoretical background. *Cem. Concr. Res.* **2001**, *31*, 647–654. [CrossRef]
16. Snoeck, D.; Pel, L.; De Belie, N. The water kinetics of superabsorbent polymers during cement hydration and internal curing visualized and studied by NMR. *Sci. Rep.* **2017**, *7*, 9514. [CrossRef]
17. De Moura, M.R.; Rubira, A.F.; Muniz, E.C. Semi-IPN Hydrogels Based on Alginate-Ca^{2+} Network and PNIPAAm: Hydrophilic, Morphological and Mechanical Properties. *Polímeros Ciência Tecnol.* **2008**, *18*, 132–137.
18. Aghaee, K.; Sposito, R.; Thienel, K.C.; Khayat, K.H. Effect of additional water or superplasticizer on key characteristics of cement paste made with superabsorbent polymer and other shrinkage mitigating materials. *Cem. Concr. Compos.* **2023**, *136*, 104893. [CrossRef]
19. Liu, J.; Farzadnia, N.; Shi, C. Microstructural and micromechanical characteristics of ultra-high performance concrete with superabsorbent polymer (SAP). *Cem. Concr. Res.* **2021**, *149*, 106560. [CrossRef]
20. Ma, X.; Liu, J.; Wu, Z.; Shi, C. Effects of SAP on the properties and pore structure of high-performance cement-based materials. *Constr. Build. Mater.* **2017**, *131*, 476–484. [CrossRef]
21. Kang, S.; Hong, S.; Moon, J. Importance of drying to control internal curing effects on field casting ultra- high performance concrete. *Cem. Concr. Res.* **2018**, *108*, 20–30. [CrossRef]
22. Liu, J.; Ou, Z.; Mo, J.; Wang, Y.; Wu, H. The effect of SCMs and SAP on the autogenous shrinkage and hydration process of RPC. *Constr. Build. Mater.* **2017**, *155*, 239–249. [CrossRef]
23. Espinoza-Hijazin, G.; Lopez, M. Extending internal curing to concrete mixtures with W/C higher than 0.42. *Constr. Build. Mater.* **2011**, *25*, 1236–1242. [CrossRef]
24. Pourjavadi, A.; Fakoorpoor, S.M.; Hosseini, P.; Khaloo, A. Interactions between superabsorbent polymers and cement-based composites incorporating colloidal silica nanoparticles. *Cem. Concr. Compos.* **2013**, *37*, 196–204. [CrossRef]
25. Justs, J.; Wyrykowski, M.; Bajare, D.; Lura, P. Internal curing by superabsorbent polymers in ultra-high performance concrete. *Cem. Concr. Res.* **2015**, *76*, 82–90. [CrossRef]
26. Xu, Z.; Yang, Q.; Zheng, X.; Liu, X.; Zhao, J.; Tan, S.H.; Liu, Z. All-weather-available down carbon fiber hydrogel with enhanced mechanical properties for simultaneous efficient clean water generation and dye adsorption from dye wastewater. *J. Colloid. Interface Sci.* **2024**, *669*, 864–876. [CrossRef]
27. Ullah, F.; Othman, M.B.H.; Javed, F.; Ahmad, Z.; Akil, H.M. Classification, processing and application of hydrogels: A review. *Mater. Sci. Eng. C* **2015**, *57*, 414–433. [CrossRef] [PubMed]
28. Bashir, S.; Hina, M.; Iqbal, J.; Rajpar, A.H.; Mujtaba, M.A.; Alghamdi, N.A.; Wageh, K.; Ramesh, K.; Ramesh, S. Fundamental Concepts of Hydrogels: Synthesis, Properties, and Their Applications. *Polymers* **2020**, *12*, 2702. [CrossRef] [PubMed]
29. Rodrigues, F.H.A.; Fajardo, A.R.; Pereira, A.G.B.; Ricardo, M.P.S.; Feitosa, J.P.A.; Muniz, E.C. Chitosan-graft-poly(acrylic acid)/rice husk ash based superabsorbent hydrogel composite: Preparation and characterization. *J. Polym. Res.* **2012**, *19*, 2012. [CrossRef]
30. Rose, V.J.O.; Sousa, C.D.; Trindade, T.M.L.S.; Barbosa, M.T.G.; Souza, N.L.G. Compósito cimentício obtido através da adição de pectina e reforçado com sílica ativa. In *A Construção Civil emu ma Perspectiva Econômica, Ambiental e Social*, 1st ed.; Araujo de Sousa, B.A., Castro, A.C., Sales, R.S., Sales, R.E.S., Eds.; Editora Científica Digital: Guarujá, Brazil, 2021; pp. 292–304.

31. Guilherme, M.R.; Aouada, F.A.; Fajardo, A.R.; Martins, A.F.; Paulino, A.T.; Davi, M.F.T.; Rubira, A.F.; Muniz, E.C. Superabsorbent hydrogels based on polysaccharides for application in agriculture as soil conditioner and nutrient carrier: A review. *Eur. Polym. J.* **2015**, *72*, 365–385. [CrossRef]
32. Mishra, R.K.; Banthia, A.K.; Majeed, A.B.A. Pectin based formulations for biomedical applications: A review. *Asian J. Pharm. Clin. Res.* **2012**, *5*, 1–7.
33. Sutar, P.B.; Mishra, R.K.; Pal, K.; Banthia, A. K Development of pH sensitive polyacrylamide grafted pectin hydrogel for controlled drug delivery system. *Int. J. Mater. Sci. Mater. Med.* **2008**, *19*, 2247–2253. [CrossRef]
34. Morello, G.; de Iaco, G.; Gigli, G.; Polini, A.; Gervaso, F. Chitosan and Pectin Hydrogels for Tissue Engineering and In Vitro Modeling. *Gels* **2023**, *9*, 132. [CrossRef]
35. Alsakhawy, M.A.; Abdelmonsif, D.A.; Haroun, M.; Sabra, S.A. Naringin-loaded Arabic gum/pectin hydrogel as a potential wound healing material. *Int. J. Biol. Macromol.* **2022**, *222*, 701–714. [CrossRef]
36. An, H.; Yang, Y.; Zhou, Z.; Bo, Y.; Wang, Y.; He, Y.; Wang, D.; Qin, J. Pectin-based injectable and biodegradable self-healing hydrogels for enhanced synergistic anticancer therapy. *Acta Biomater.* **2021**, *131*, 149–161. [CrossRef]
37. Rúfin-Gómez, L.A.; Delgafo-Pérez, L.G.; Méndez-Martínez, J. Efectos patogénicos de la acrilamida para la salud. Una revision. *Rev. Méd. Electrón.* **2022**, *44*, 388–402.
38. Bušová, M.; Bencko, V.; Laktičová, K.V.; Holcátová, I.; Vargová, M. Risk of exposure to acrylamide. *Cent. Eur. J. Public. Health* **2020**, *28*, S43–S46. [CrossRef]
39. Sankar, S.; Sharma, S.K.; Kaur, N.; Lee, B.; Kim, D.Y.; Lee, S.; Jung, H. Biogenerated silica nanoparticles synthesized from sticky, red, and brown rice husk ashes by a chemical method. *Ceram. Int.* **2016**, *42*, 4875–4885. [CrossRef]
40. Eliche-Quesada, D.; Felipe-Sesé, M.A.; López-Pérez, J.A.; Infantes-Molina, A. Characterization and evaluation of rice husk ash and wood ash in sustainable clay matrix bricks. *Ceram. Int.* **2017**, *43*, 463–475. [CrossRef]
41. Cusson, D.; Hoogeveen, T. Internal curing of high-performance concrete with pre-soaked fine lightweight aggregate for prevention of autogenous shrinkage cracking. *Cem. Concr. Res.* **2008**, *38*, 757–765. [CrossRef]
42. Bezerra, I.M.T.; Souza, J.; Carvalho, J.B.Q.; Neves, G.A. Application of the rice husk ash in mortars for bricklaying. *R. Bras. Eng. Agríc. Ambiental.* **2011**, *15*, 639–645. [CrossRef]
43. Brandão Júnior, J.R.; Lopes, F.F.M.; Oliveira, H.C.M.; Lopes Neto, J.P.; Silvino, G.S. Characterization of red rice shell ash for application as pozzolanic material. *Energ. Agric.* **2020**, *35*, 80–88.
44. Associação Brasileira de Normas Técnicas. *NBR 7214: Areia Normal para Ensaio de Cimento—Especificação*; ABNT: Rio de Janeiro, Brazil, 2015.
45. Cândido, J.S.; Leitão, R.C.F.; Ricardo, N.M.P.S.; Feitosa, J.P.A.; Muniz, E.C.; Rodrigues, F.H.A. Hydrogels composite of poly(acrylamide-co-acrylate) and rice husk ash. I. Synthesis and characterization. *J. Appl. Polym. Sci.* **2011**, *123*, 879–887. [CrossRef]
46. Associação Brasileira de Normas Técnicas. *NBR 7215: Cimento Portland—Resistência à Compressão do Cimento Portland*; ABNT: Rio de Janeiro, Brazil, 1997.
47. Cândido, J.S.; Pereira, A.G.B.; Fajardo, A.R.; Ricardo, N.M.P.S.; Feitosa, J.P.A.; Muniz, E.C.; Rodrigues, F.H.A. Poly(acrylamide-co-acrylate)/rice husk ash hydrogel composites. II. Temperature effect on rice husk ash obtention. *Compos. B Eng.* **2013**, *51*, 246–253.
48. Spagnol, C.; Rodrigues, F.H.A.; Pereira, A.G.B.; Fajardo, A.R.; Rubira, A.F.; Muniz, E.C. Superabsorbent hydrogel nanocomposites based on starch-g-poly(sodium acrylate) matrix filled with cellulose nanowhiskers. *Cellulose* **2012**, *19*, 1225–1273. [CrossRef]
49. Gharekhani, H.; Olad, A.; Mirmoshseni, A.; Bybordi, A. Superabsorbent hydrogel made of NaAlg-g-poly(AA-co-AAm) and rice husk ash: Synthesis, characterization, and swelling kinetic studies. *Carbohydr. Polym.* **2017**, *168*, 1–13. [CrossRef]
50. El-Hamshary, H. Synthesis and water sorption studies of pH sensitive poly(acrylamide-co-itaconic acid) hydrogels. *Eur. Polym. J.* **2007**, *43*, 4830–4838. [CrossRef]
51. Garcia, J.A.F.; de Moura, M.R.; Aouada, F.A. Effect of pH, ionic concentration and species on the absorption of water by hydrogel bionanocomposits constituted from CMC/PAAm/LAPONITE RDS. *Quim. Nova* **2019**, *42*, 831–837.
52. Spagnol, C.; Rodrigues, F.H.A.; Pereira, A.G.B.; Fajardo, A.R.; Rubira, A.F.; Muniz, E.C. Superabsorbent hydrogel composite made of cellulose nanofibrils and chitosan-graft-poly(acrylic acid). *Carbohydr. Polym.* **2013**, *87*, 2038–2045. [CrossRef]
53. Spagnol, C.; Rodrigues, F.H.A.; Neto, A.G.V.C.; Pereira, A.G.B.; Fajardo, A.R.; Radovanovic, E.; Rubira, A.F.; Muniz, E.C. Nanocomposites based on poly(acrylamide-co-acrylate) and cellulose nanowhisker. *Eur. Polym. J.* **2012**, *48*, 454–463. [CrossRef]
54. Chandrasekhar, S.; Pramada, P.N.; Praveen, L. Effect of organic acid treatment on the properties of rice husk silica. *J. Mater. Sci.* **2005**, *40*, 6535–6544. [CrossRef]
55. Tanwar, A.; Date, P.; Ottoor, D. ZnO NPs incorporated gelatin grafted polyacrylamide hydrogel nanocomposite for controlled release of ciprofloxacin. *J. Colloid. Interface Sci.* **2021**, *42*, 100413. [CrossRef]
56. Kumar, M.; Mishra, R.K.; Banthia, A.K. Development of pectin based hydrogel membranes for biomedical applications. *Int. J. Plast. Technol.* **2010**, *14*, 213–223. [CrossRef]
57. Khan, Z.; Minhas, M.U.; Ahmad, M.; Khan, K.U.; Sohail, M.; Khalid, I. Functionalized pectin hydrogels by cross-linking with monomer: Synthesis, characterization, drug release and pectinase degradation studies. *Polim. Bull.* **2020**, *77*, 339–356. [CrossRef]

58. Pandey, M.; Choudhury, H.; D/O Singh, S.K.S.; Annan, N.C.; Bhattamisra, S.K.; Gorain, B.; Amin, M.C.I.M. Budesonide-Loaded Pectin/Polyacrylamide Hydrogel for Sustained Delivery: Fabrication, Characterization and In Vitro Release Kinetics. *Molecules* **2021**, *26*, 2704. [CrossRef]
59. Danuté, V.; Gintautas, S.; Mindaugas, D.; Vytautas, S. Cement hydration with zeolite-based additive. *Chemija* **2013**, *24*, 271–278.
60. Putra, I.M.W.A.; Arryanto, Y.; Kartini, I. Preparation of cement-zeolite-clay composite plate and its adsorption-desorption study with respect to ethanol/water mixture. In Proceedings of the 2nd International Conference on Chemical Sciences Proceeding, Yogyakarta, Indonesia, 14–16 October 2010.
61. Araújo, A.G., Jr.; Vale, A.E.; Azevedo, A.G.S.; Strecker, K. Study of recycling Portland cement of high initial strength. *Cerâmica* **2017**, *63*, 58–64.
62. Lin, R.S.; Wang, X.Y.; Lee, H.S.; Cho, H.K. Hydration and Microstructure of Cement Pastes with Calcined Hwangtoh Clay. *Materials* **2019**, *12*, 458. [CrossRef] [PubMed]
63. Biricik, H.; Sarier, N. Comparative Study of the Characteristics of Nano Silica–, Silica Fume– and Fly Ash–Incorporated Cement Mortars. *Mat. Res.* **2014**, *17*, 570–582. [CrossRef]
64. Ghoddousi, P.; Saadabadi, L.A. Study on hydration products by electrical resistivity for self-compacting concrete with silica fume and metakaolin. *Constr. Build. Mater.* **2017**, *154*, 219–228. [CrossRef]
65. Abo-El-Enein, S.A.; Heikal, M.; Amin, M.S.; Negm, H.H. Reactivity of dealuminated kaolin and burnt kaolin using cement kiln dust or hydrated lime as activators. *Constr. Build. Mater.* **2013**, *47*, 1451–1460. [CrossRef]
66. Tavares, A.M.; Andrade, J.; Cestari, A.R.; Vieira, E.F. Synthesis and characterization of cement slurries additives with epoxy resins-kinetics, thermodynamic and calorimetric analysis. In Proceedings of the 19th Brazilian Congress of Materials Science and Engineering—CBECiMat, Campos do Jordão, Brazil, 21–25 November 2010.
67. Silva, L.A.; Nahime, B.O.; Lima, E.C.; Akasaki, J.L.; Reis, I.C. XRD investigation of cement pastes incorporating concrete floor polishing waste. *Cerâmica* **2020**, *66*, 373–378. [CrossRef]
68. Rupasinghe, M.; Nicolas, R.S.; Mendis, P.; Sofi, M.; Ngo, T. Investigation of strength and hydration characteristics in nano-silica incorporated cement paste. *Cem. Concr. Compos.* **2017**, *80*, 17–30. [CrossRef]
69. Prabahar, J.; Vafaei, B.; Ghahremaninezhad, A. The Effect of Hydrogels with Different Chemical Compositions on the Behavior of Alkali-Activated Slag Pastes. *Gels* **2022**, *8*, 731. [CrossRef]
70. Meng, S.; Ouyang, X.; Fu, J.; Niu, Y.; Ma, Y. The role of graphene/graphene oxide in cement hydration. *Nanotechnol. Rev.* **2021**, *10*, 768–778. [CrossRef]
71. Santos, T.A.; de Oliveira e Silva, G.A.; Ribeiro, D.V. Mineralogical Analysis of Portland Cement Pastes Rehydrated. *JSWTM* **2020**, *46*, 15–23. [CrossRef]
72. Baloch, H.; Usman, M.; Rizwan, S.A.; Hanif, A. Properties enhancement of super absorbent polymer (SAP) incorporated self-compacting cement pastes modified by nano silica (NS) addition. *Constr. Build. Mater.* **2019**, *203*, 18–26. [CrossRef]
73. Schröfl, C.; Mechtcherine, V.; Gorges, M. Relation between the molecular structure and the efficiency of superabsorbent polymers (SAP) as concrete admixture to mitigate autogenous shrinkage. *Cem. Concr. Res.* **2012**, *42*, 865–873. [CrossRef]
74. He, R.; Tan, Y.; Chen, H.; Wang, Z.; Zhang, J.; Fang, J. Preparation and properties of novel superabsorbent polymer (SAP) composites for cementitious materials based on modified metakaolin. *Constr. Build. Mater.* **2020**, *258*, 119575. [CrossRef]

materials

MDPI

rticle

’olyacrylamide Hydrogel Containing Starch and Sugarcane ;agasse Ash: Synthesis, Characterisation, and Application in :ement Pastes and Mortars

ına Elizabete Nunes Pereira [1], Edson Araujo de Almeida [2], Fábio Rodrigo Kruger [3], dson Cavalcanti da Silva-Filho [1] and Edvani Curti Muniz [1,2,4,*]

1 Chemistry Postgraduation Program, Federal University of Piaui—UFPI, Teresina 64049-550, PI, Brazil; anaelizabete.np@gmail.com (A.E.N.P.); edsonfilho@ufpi.edu.br (E.C.d.S.-F.)
2 Chemistry Postgraduation Program, State University of Maringa—UEM, Maringá 87020-900, PR, Brazil; edsonalmeida2009@gmail.com
3 Civil Construction Department, Federal University of Technology-Parana—UTFPR-CM, Campo Mourão 87301-889, PR, Brazil; fabiorodrigo_civil@hotmail.com
4 Materials Science Postgraduation Program, Federal University of Technology-Parana—UTFPR-LD, Londrina 86036-370, PR, Brazil
* Correspondence: munizec@ufpi.edu.br

Abstract: Internal curing is a process based on the addition of materials that function as water reservoirs in cementitious media. Superabsorbent hydrogels are an alternative that can be used as an internal curing agent, as they have the ability to absorb and release water in a controlled manner. In the present work, superabsorbent hydrogels based on crosslinked polyacrylamide in the presence of starch and sugarcane bagasse ash (SCBA) were developed and applied to mortars as an internal curing agent. The synthesized hydrogels were evaluated by SEM, FTIR, and swelling analysis. Cement pastes and mortars were produced using different amounts of hydrogel (0.03%, 0.06%, and 0.1% by weight). An analysis of the cement pastes and mortars revealed that hydrogel contributes to hydration, thus improving the quality of the product. Furthermore, the addition of 0.03% hydrogel by weight increased the mechanical resistance of the mortars in up to 26.8% at 28 days of curing as compared with reference (without hydrogel). To the best of our knowledge, this is the first study to use a hydrogel based on polyacrylamide crosslinked with starch and SCBA as a curing agent for mortars and cement pastes. This approach is environmentally friendly, because it uses a natural product (starch) and a byproduct from the sugarcane industry (SCBA).

Keywords: hydrogel; starch; sugarcane bagasse; hydrolysis; mortar; mechanical resistance

1. Introduction

In recent decades, there has been a marked growth in both civil construction and the trailer industry. These changes have demanded new types of cementitious materials that meet different requests in current and unconventional applications [1]. One of the most used materials in this field is cement, which can be utilised in the production of concrete and mortars [2]. Mortars are heterogeneous composite materials used in civil construction; they comprise a binder (cement, lime, or gypsum), fine aggregate (sand), and water. Their use is versatile: they can serve as binding elements for other materials (blocks, bricks, and stones) and be used as surface coverings, such as plaster and base for ceramic coverings [3].

When a cementitious material comes into contact with water, hydration reactions begin and underlie the curing process. Proper mortar curing requires maintaining a satisfactory moisture content and temperature in freshly cast mortar for a specific period of time immediately following placement [4]. An effective curing process can drastically reduce the appearance of fissures and cracks [5]. So, it is important to maintain constant moisture in the mix during mortar curing, which minimises the effect of water loss through evaporation. Curing also helps maintain heat in the mixture, which speeds up the chemical reaction between cement and water [6]. In the case of cement, the cementitious mixture is moistened by the presence (or further introduction) of water for the hydration reactions. The main product of cement hydration is calcium silicate hydrate (C-S-H). Poorly cured concrete loses around 30% of its strength, thus facilitating the appearance of fissures and cracks in its structure [7].

There has been much attention directed towards the curing process due to the technological importance of cementitious materials, including their expected long life and the safety of their use [8]. Currently, internal curing is gaining prominence as an alternative to improve the hydration of the cementitious environment [9]. Internal curing involves supplying water to the aggregates in the cementitious material, which can occur by incorporating a curing agent into the cement and/or mortars. In other words, as the hydration reactions take place, the internal curing agent releases water in a controlled manner into the cementitious matrix [9,10]. The released water is concomitantly used in the chemical reactions that comprise the curing process.

One of the options to increase the water concentration in a cementitious material is the application of superabsorbent hydrogels [11]. Hydrogels are networks of crosslinked polymers that can absorb large amounts of water and release the absorbed water in a gradated manner over time. The water inside the hydrogel structure allows for the diffusion of the solute through the swollen polymer, allowing it to be applied as a release system control [12–14]. In this sense, hydrogels participate in the controlled release of water and can improve the continuation of hydration reactions that occur over time in the cementitious matrix, thus ensuring a more satisfactory internal cure [3]. Hydrogels can be synthesised from different materials, using synthetic ones, such as acrylamide/acrylate monomers, and/or natural polymers, such as polysaccharides [15]. A good example of a natural polymer used in hydrogel formulations is starch due to its composition, structure, and great abundance [15,16]. Starch is a complex polysaccharide composed of two components, amylose and amylopectin, each of which presents a specific characteristic structure [15,16]. Starch appears in a considerable proportion in plants, including cereal grains (rice, corn, etc.) and tubers (potatoes, cassava, etc.). So, it is easy to acquire in natural or chemically modified form [17], and it is often not expensive. Therefore, starch is one of the important natural polymeric materials [16,17].

Hydrogels prepared through the combination of natural and synthetic moieties have been described in the literature [15,18] and represent a good alternative for several technological applications [13,15,19–21]. The swelling process of a hydrogel is guided by physical factors related to the three-dimensional (3D) network, such as the degree of crosslinking and the presence of hydrophilic groups (e.g., -COOH, -OH, and-NH_2) in the structure of the polymeric chains, and by external factors, such as the temperature, pH, and ionic strength [12,20,21]. The presence of hydrophilic groups on polymeric chains contributes to lowering the required crosslinking density and allows for high flexibility of the polymeric network [22]. In this sense, low crosslinking and high chain flexibility certainly contribute to the swelling of the studied material. The incorporation of sugarcane bagasse ash (SCBA), a burned agro-industrial residue, into the formulations for a cementitious matrix is another alternative to improve the properties of cement-based materials (i.e., concrete and mortars). The amount of silica in SCBA can facilitate curing and might improve the mechanical properties of the prepared concrete and mortar [23,24].

This work presents a new approach to prepare hydrogels, namely polymerisation/crosslinking of acrylamide in the presence of starch and SCBA. The hypothesis is that applying such a hydrogel in mortar and concrete improves the mechanical properties [23,24]. So, mortar formulations with different hydrogel content (0.03, 0.06, and 0.10 by weight) were prepared. The prepared hydrogels were characterised with scanning electron microscopy, Fourier transform infrared (FTIR) spectroscopy, and water absorption/desorption analysis, and the mortars were characterised by FTIR spectroscopy, X-ray diffraction (XRD), thermogravimetric analysis (TGA), and compressive strength assays.

2. Materials and Methods

2.1. Materials

Acrylamide (AAm, Sigma-Aldrich, St. Louis, MO, USA) was used as a monomer in colourless solid form with a melting point of 80 °C, boiling point of 125 °C, and density of 1.12 g/cm^3; potassium persulfate (Synth, São Paulo, Brazil) in white powder form, with a melting point of 100 °C and density of 2.477 g/cm^3, was used an as initiator; and *N,N′*-methylenebisacrylamide (MBAAm, Fisher Scientific, Hampton, NH, USA) in white powder form, with a melting point above 300 °C, boiling point of 333.8 °C, and density of 1.216 g/cm^3, at 20 °C was used as a crosslinking agent. The hydrogels also comprise commercial cassava starch (dissolved in distilled water) and SCBA; the properties of SCBA may vary depending on the conditions in which the sugar cane was grown. The SCB was purchased locally in Teresina–Piauí, Brazil and then burned at a specific temperature at atmospheric pressure to obtain SCBA. The mortars were prepared with Portland cement type CP II-F and sand (normal Brazilian in four fractions: coarse (1.19 mm), medium coarse (0.59 mm), medium fine (0.297 mm), and thin (0.149 mm), according to the ABNT/NBR 7214/2015 [25].

2.2. Methods

2.2.1. Preparation of SCBA, Synthesis, and Hydrolysis of the Hydrogels

The SCBA was obtained by burning SCB in an EDG-3000 muffle furnace at 600 °C for 2 h. Subsequently, the ash was crushed and sieved (75 mesh) [1]. The hydrogels were prepared by using a 2^3 factorial design with three replications at the central point, totalling 11 formulations. The influence of three factors—(i) the amount of acrylamide, (ii) the amount of starch, and (iii) the amount of SCBA—(shown in Table 1) on the swelling capacity of the hydrogels was analysed. The levels for each factor were defined based on studies that have already synthesized acrylamide hydrogels in the presence of rice husk ash [26]. From this factorial design (shown in Table 2), the hydrogels with the highest swelling values were chosen for the subsequent experiments.

Table 1. Factors and their levels used in the development of factorial design.

Factors (g)	Lower Level (−)	Central Point (0)	Upper Level (+)
AAm	1.5	2.5	3.5
Starch	1.0	2.0	3.0
SCBA	0.10	0.20	0.30

Table 2. Complete factorial design matrix with all formulations used for hydrogel synthesis.

Run	AAm Content	Starch Content	SCBA Content
H1	1.5	1.0	0.10
H2	3.5	1.0	0.10
H3	1.5	3.0	0.10
H4	3.5	3.0	0.10
H5	1.5	1.0	0.30
H6	3.5	1.0	0.30
H7	1.5	3.0	0.30
H8	3.5	3.0	0.30
H9a	2.5	2.0	0.20
H9b	2.5	2.0	0.20
H9c	2.5	2.0	0.20

To prepare the hydrogels, initially, the starch was dissolved in distilled water (using sufficient volume to maintain the concentration of the entire formulation at 136 g L^{-1}) at 50 °C and magnetically stirred until its complete solubilization. Then, the solution was stirred for 24 h at room temperature. After this process, the required amount of SCBA was added, keeping the system at 70 °C with magnetic stirring. After 10 min, the required amount of AAm, 0.04 g of MBAAm, as a crosslinker, and 0.08 g of potassium persulfate, as an initiator, were added, with 10 min between the addition of each reactant. The system was stirred until it gelled. The hydrogels were cut into small pieces and immersed in distilled water (ca. 1 g swollen hydrogel per 50 mL of water) for washing. The water was changed every hour for 6 h. Then, the washed material was placed in an oven at 60 °C for 24 h to dry. After drying, the hydrogels were crushed and sieved (250 mesh).

For the hydrolysis of the hydrogels, 40 mL of 0.5 mol L^{-1} NaOH solution was added per 1 g of dry hydrogel. In a hood, the solution was magnetically stirred for 1 h at 50 °C. Then, the hydrogels were washed with distilled water until reaching pH 7 and, subsequently, dried in an oven at 60 °C for 24 h [27].

2.2.2. Preparation of Mortars and Cement Pastes

The hydrolysed H2 hydrogel (Table 2) was applied to the mortar formulations at 0.03%, 0.06%, or 0.1% relative to the mass of the cement (*w*/*w*). The hydrolysed H2 hydrogel was chosen due to its degree of swelling (Q) of 200.20 g/g. The mortars were prepared in accordance with ABNT/NBR 7215 standards [28]. The cement/sand ratio was 1:3. As described in Section 2.1, four fractions of sand were used. The hydrogels were crushed, sieved (250 mesh), and swollen over a period of 20 h. The cement pastes were produced by applying the same amount of hydrolysed H2 hydrogel used in mortars, with the aim of better understanding the hydration process, because they are less complex than mortars, as they only contain cement hydration products [29,30]. Figure 1 shows the preparation of the mortars and cement pastes.

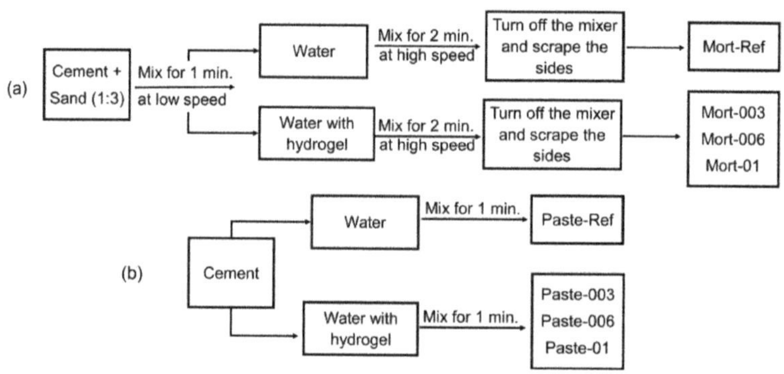

Figure 1. Procedure for preparing the (**a**) mortar and (**b**) cement paste formulations.

2.3. Characterisation

2.3.1. Degree of Swelling and Desorption of Hydrogels in Aqueous Solutions

The water absorption of the prepared hydrogels over time was measured using solutions at pH 4, 7, and 11 and in saline solution (0.1 mol L^{-1} NaCl). For these measurements, approximately 10 mg of dry hydrogel was used. Each hydrogel was immersed in 100 mL of solution and weighed with an analytical balance after 1, 2, 3, 4, 5, 6, 7, 8, 24, 48, and 72 h. The degree of swelling (Q) was calculated using Equation (1):

$$Q = [(m - m_0)/m_0] \quad (1)$$

where m is the mass of the swollen hydrogel, and m_0 is the dry mass of the hydrogel sample. To measure desorption, the hydrogels were swollen to equilibrium in distilled water and then transferred to an oven at 40 °C. The mass was determined after 1, 2, 3, 4, 5, 6, and 24 h. The results are represented as the percentage of remaining water, with the initial mass of the swollen hydrogel considered to be 100% water. The swelling and desorption tests were carried out in triplicate.

2.3.2. Fourier Transform Infrared Spectroscopy (FTIR)

The hydrogel samples were dried and crushed with 250 mesh. Attenuated total reflectance (ATR)-FTIR spectra were obtained using a Vertex 70 apparatus (Bruker, Billerica, MA, USA) in the range of 600–4000 cm^{-1} for the hydrogels and the range of 400–4000 cm^{-1} for the mortars. The mortars and cement pastes were aged 7 and 28 days and then dried and crushed. Pastilles were prepared in the form of potassium bromide tablets at 1% (*w*/*w*).

2.3.3. X-Ray Diffraction (XRD)

X-ray diffractograms of the mortars, in powder form, for 2(θ) of 10°–70° were obtained using an XRD-6000 diffractometer (Shimadzu, Kyoto, Japan) operating at a speed of 1° min^{-1}, a divergent angle of 1°, 40 kV Cu radiation, and a current of 30 mA.

2.3.4. Thermogravimetric (TGA)

Thermogravimetric was performed to monitor the degree of hydration and curing reactions of mortars aged for 28 days. A DTG-60 apparatus (Shimadzu) was used with nitrogen as a carrier gas, a temperature of 25–600 °C, a heating rate of 10 °C, and a flow rate of 100 mL min^{-1}.

2.3.5. Scanning Electron Microscopy (SEM)

A Quanta 250 scanning electron microscope (FEI) was used to analyse the morphology of the hydrogels. The samples were covered with a thin layer of gold (ca. 50 nm). The microscope operated at an accelerating voltage of 20 kV. The pore size and its distribution were measured using Image J® 1.8.0 software; for this, the pore contours were manually delimited through the SEM images.

2.3.6. Compressive Strength Tests

The compressive strength tests of the mortars were performed according to the NBR 7215 standard (ABNT/1996) [28,31], using cylindrical specimens with 50 mm in diameter and 100 mm in height. The samples were aged for 7 and 28 days and analyzed in quadruplicate in an EMIC Pc 100 machine, applying a load of 0.3–0.8 MPa s^{-1} until rupture. The tests provided the force applied to the specimen. The individual strength was calculated in mega Pascals (MPa) as a function of the specimen cross area.

3. Results

3.1. Water Absorption of Hydrogels

Figure 2a shows the degree of swelling in distilled water of the non-hydrolysed hydrogels as a function of hydration time. Figure 2b demonstrates that amide hydrolysis in an alkaline medium effectively increased the degree of swelling of the hydrogels more than 10-fold compared with the non-hydrolysed ones. Moreover, both non-hydrolysed and hydrolysed hydrogels showed a variable degree of swelling, probably due to the amount of AAm used in the formulation. The swelling equilibrium time was around 20 h for the non-hydrolysed hydrogels (Figure 2a) and >50 h for the hydrolysed hydrogels (Figure 2b). The presence of hydrophilic and hydrophobic groups is another factor linked to the absorption of water by the hydrogel, since hydrophilic groups interact with water, causing it to remain inside the hydrogel, unlike hydrophobic groups that prevent water from remaining inside the hydrogel. The synthesized hydrogels have amide -NH_2 groups, hydroxyls OH, and also, the presence of silanol groups (SiO_2) that can interact with water, causing it to remain inside the hydrogel.

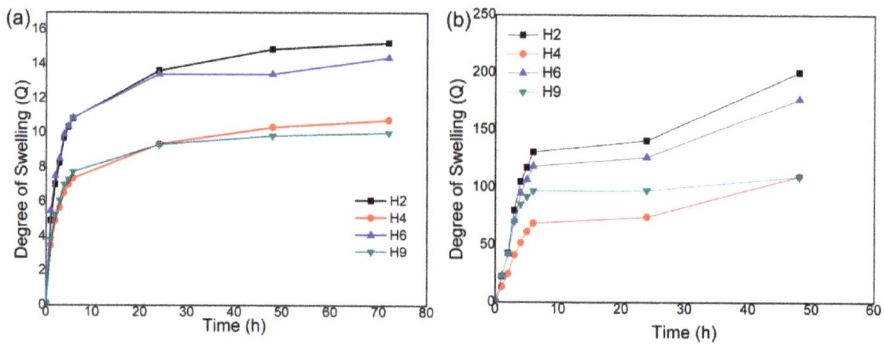

Figure 2. The degree of swelling as a function of time for the (**a**) non-hydrolysed and (**b**) hydrolysed H2, H4, and H9 hydrogels.

Non-hydrolysed polyacrylamide hydrogels undergo changes from neutral hydrogels (before hydrolysis) to partially charged hydrogels (polyelectrolytes) after hydrolysis. As the neutral polyacrylamide hydrogel is immersed in an alkaline solution for a given time, the amide groups (-$CONH_2$) are gradually transformed into ionized carboxyl groups due to hydrolysis. In other words, the amide groups in the polymeric chains in the network react with hydroxyl ions in an alkaline solution and are partially converted into the carboxylate and amine groups, [-$CONH_2$ + NaOH $\rightarrow$ -COO^-Na^+ + NH_3]. Therefore, the hydrolysed polyacrylamide/starch/SCBA in the H2 hydrogel increased its volume by almost 200-fold compared with the dry hydrogel, much higher than the 15-fold volume increase in the non-hydrolysed H2 hydrogel. This is due to the repulsion of negatively charged carboxylate groups in the hydrolysed hydrogel. Carboxylic groups interact more with water, thus leading to a higher degree of swelling [32–35].

PAAm hydrogels resulting from the polymerization of acrylamide present lower swelling values (Q) when the hydrogel presents a high degree of crosslinking, because the more crosslinked the polymer chain is, the less flexibility it will have, which makes it difficult for the expansion to occur and for water to diffuse into the interior of the hydrogel matrix. It can be noted that the hydrogels before the hydrolysis process presented swelling values of 10–15 g/g, which is related to the crosslinking of the polymer chain; that is, the polymer chain was not very flexible, which influenced the diffusion of water into the interior of the hydrogel. After hydrolysis, the formation of -COO^- groups occurred, which undergo electrostatic repulsion between themselves, being capable of increasing the expansion of the polymer chain and, thus, increasing the diffusion of water into the interior of the hydrogel. Consistently, the hydrolysed hydrogels presented a higher degree of swelling than the non-hydrolysed hydrogels. This fact arises from the interaction between the hydrogel that is favoured by hydrolysis. However, this interaction is not necessarily fast.

3.2. *Swelling Behaviour of the Hydrogels at Different pHs and in Saline Solution*

Figure 3 shows the degree of swelling of the hydrolysed hydrogels at different pHs. The hydrogels, except for the H2 hydrogel, swelled less in the acidic medium compared with the neutral and alkaline media. In this scenario, the carboxylate groups undergo protonation in acidic medium (pKa $\approx$ 4.5), and this minimises the anion–anion repulsive forces [36]. This fact may be related to the ionisation of the carboxyl groups present in the hydrogels. In this case, ionisation promotes the formation of carboxylate and through electrostatic repulsion between the carboxylate-containing chains, promoting the diffusion of water into 3D matrix to minimise the repulsive forces, leading to an expansion of the entire polymeric chain's matrix [27]. The hydrogel becomes less hydrophilic when the pH is less than the pKa, which decreases the degree of swelling. Hydrogen bonds are possible in an acidic environment; that is, the degree of crosslinking of the chains increases.

Therefore, the formation of hydrogen bonds hinders the movement of the chains in the hydrogel, forming a consistent hydrogel network when the pH is less than the pKa [27,37].

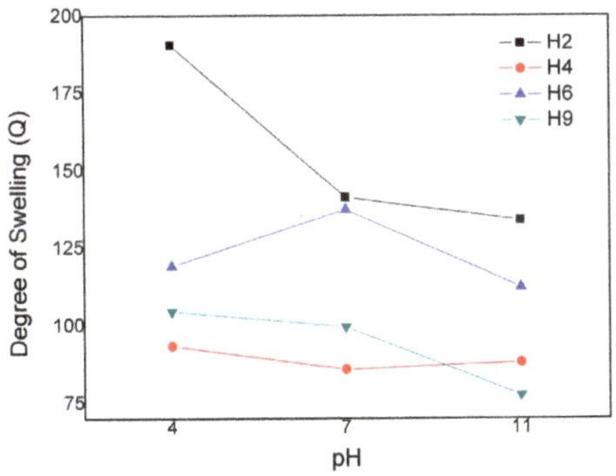

Figure 3. Degree of swelling of the hydrolysed H2, H4, H6, and H9 hydrogels in acidic (pH 4), neutral (pH 7.0), and alkaline (pH 11) media.

The degree of swelling increased for the H6 and H9 hydrogels, but not for the H2 and H4 hydrogels, at pH 7 compared with pH 4. The H4 hydrogel presented the smallest Q value at this pH. The H4 hydrogel contains more starch (approximately 3.5 g) than the H2 (1.0 g), H6 (1.0 g), and H9 (2.0 g) hydrogels. The higher amount of starch reduced the Q value at pH 7 due to increased mechanical crosslinking (entanglement) points in the polymeric network, which increased the crosslinking density of the hydrogel, resulting in a decrease in water diffusion in the hydrogel. This decrease may also be related to the fact that starch is partially filled in the hydrogel network; therefore, the number of hydrophilic groups is considerably smaller, and the Q value tends to decrease [38].

At pH 10, the Q value increased for the H4 hydrogel and decreased for the H2, H6, and H9 hydrogels. This phenomenon may be related to the ionisation of the carboxy groups present in the studied hydrogel formulations. Ionisation promotes electrostatic repulsion between the chains, promoting the diffusion of water into the 3D matrix to minimise the repulsive forces, thus expanding the entire polymeric chain matrix [27].

In summary, the H2 and H6 hydrogels presented the highest Q values and the H4 and H9 hydrogels presented the lowest Q values in all studied media. These data may be related to the starch content present in the gels, given that the H2 and H6 hydrogels have the lowest starch content, while the H4 and H9 hydrogels have the highest starch content. The data also indicate that the addition of SCBA does not influence the swelling process as much as starch: the H6 hydrogel has the highest SCBA content, but its Q values were higher than the H4 and H9 hydrogels, which have the lowest SCBA content.

Due to the large number of ions present in the cement matrix, it is important to evaluate the performance of the degree of swelling in saline solution, which is representative of the cement medium. Figure 4a shows the Q values for the H2, H4, H6, and H9 hydrogels in saline solution. The H2 hydrogel presented the highest Q value, and the H9 hydrogel presented the lowest Q value. As shown in Figure 4b, the Q values in saline solution reduced sharply compared with acidic, neutral, and alkaline media, as denoted by the sensitivity factor (f), calculated with Equation (2):

$$f = 1 - [Qsal/Qwater] \quad (2)$$

where Qsal is the degree of swelling in saline solution, and Qwater in distilled water. Based on the f values, the saline effects are not as pronounced for the H9 and H4 hydrogels, as they are from the H2 and H6 hydrogels. The results indicate that the ash content does not affect the f value. Ash calcined at 600 °C was used in all the hydrogels: 10% for the H2 and H4 hydrogels, 30% for the H6 hydrogel, and 20% for the H9 hydrogel. The amount of

SCBA in hydrogels and the ash granulometry should affect the swelling capacity. It was expected that, in high quantities, ash can fill the pores of the gels, making it difficult to increase Q. However, this behaviour was not observed for the solution, because the H6 hydrogel presented higher Q values than the H4 and H9 hydrogels.

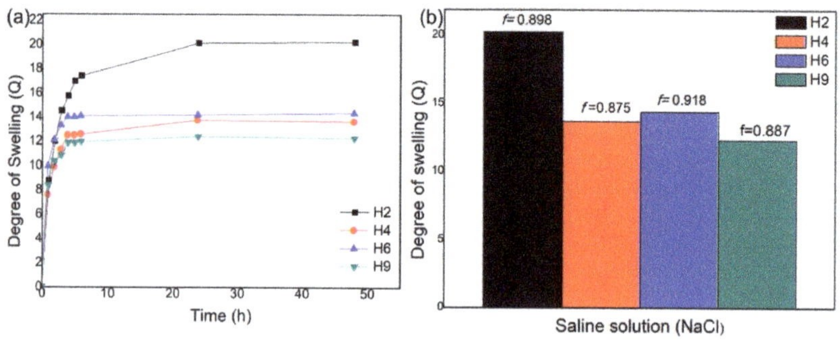

Figure 4. (**a**) A graph showing the degree of swelling graph as a function of time in saline solution for the non-hydrolysed H2, H4, H6, and H9 hydrogels; (**b**) the sensitivity factor (f) for the H2, H4, H6, and H9 hydrogels at equilibrium.

These results corroborate findings from the literature: the presence of ions reduces the degree of swelling [24]. When the hydrogel is immersed in a saline solution (containing positive ions), interactions may occur between the hydroxyl groups present in the polymer chain and the sodium ions of NaCl, thus forming ionic pairs between the species [39,40]. In other words, there are electrostatic repulsions between the polymer chains of the hydrogel [22,41]. The contraction of polymer networks may also occur, contributing to reduce the hydrophilicity of the material.

3.3. Evaluation of Desorption of Water Absorbed by the Hydrolysed Hydrogels

Figure 5 shows the results of the water desorption assays. For this analysis, the swelling achieved during immersion for 24 h was considered to be the initial swelling. Therefore, the initial mass indicates 100% water present in the hydrogel matrix. Within 6 h, the H4 and H6 hydrogels released around 50% of the absorbed water, while the H2 and H9 hydrogels released 35% of the absorbed water. Furthermore, by 24 h, each hydrogel had released around 95% of the absorbed water, with no difference between the hydrogels. The desorption test was carried out in water, but the results are expected to be different in mortar. Therefore, if the hydrogel releases all the water retained in the polymeric matrix within 24 h, then the hydrogel is not completely effective, because mortar cures over a period of more than 28 days [4].

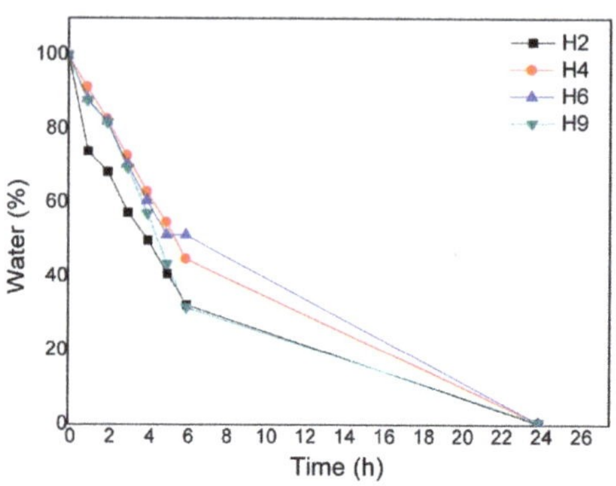

Figure 5. Water desorption from the H2, H4, H6, and H9 hydrogels over time.

3.4. Morphology of the Hydrogels

Figure 6a–d shows representative scanning electron micrographs of the hydrolysed H2, H4, H6, and H9 hydrogels. They show that the hydrogels presented porous surfaces, as pores in irregular shapes and sizes are distributed throughout the polymer matrix, thus indicating that the addition of SCBA did not prevent the formation of pores in the hydrogels. Figure 6e–h shows the average pore size of the hydrolysed H2, H4, H6, and H9 hydrogels: 4804.97 ± 2067.45, 2772.35 ± 1130.99, 5159.91 ± 3759.45, and 4460.17 ± 1838.54 μm, respectively. It can be noted that the presence of SCBA can influence the formation of pores, since the H6 hydrogel presented larger pores than the others, and its formulation has the highest SCBA content. The H6 hydrogel has a larger pore size compared with the other hydrogels, a surprising finding given that it did not have the highest Q value; this may be related to a low starch content (1.0 g) compared with the H4 and H9 hydrogels (3.0 g and 2.0 g, respectively), thus indicating that the interactions between the hydrophilic groups present in the polymer matrix favour swelling more than the hydrogel pore size, considering that the H4 and H9 hydrogels have smaller pore sizes and higher Q values than the H6 hydrogel.

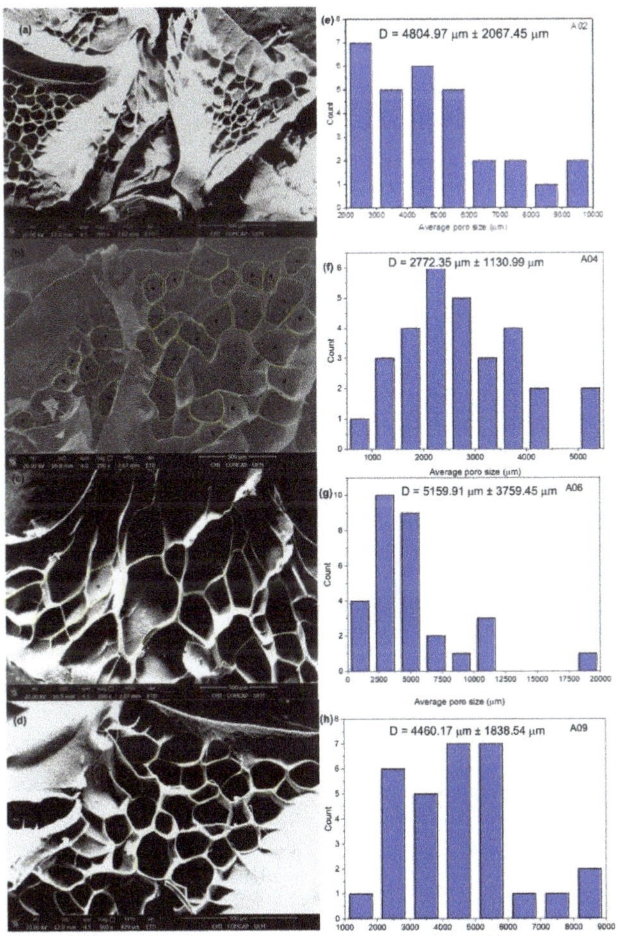

Figure 6. Scanning electrons micrographs (200× magnification) and the pore distribution of the hydrolysed (**a**) and (**e**) H2, (**b**) and (**f**) H4, (**c**) and (**g**) H6, and (**d**) and (**h**) H9 hydrogels.

3.5. FTIR of the Hydrogels and XRD of SCBA

The FTIR spectra of SCBA, PAAm, starch, and the hydrolysed H2, H4, H6, and H9 hydrogels are shown in Figure 7. For SCBA, the bands at 3392 and 1664 cm^{-1} are attributed to the O-H deformation of the silanol group (-SiOH); the bands at 1043 and 667 cm^{-1} are attributed, respectively, to the symmetric stretching and deformation of Si-O-Si [40,42]. For PAAm (formed from the polymerization of acrylamide (Amm)), the bands at 3357 and 3195 cm^{-1} refer to the axial deformation of the N-H bond, and the bands at 1676 cm^{-1} and 1605 cm^{-1} are attributed to the stretching vibration of the C=O groups and the N-H stretching vibration of the amide groups, respectively, belonging to the PAAm molecule [43,44]. For starch, the band at 3435 cm^{-1} refers to the O–H stretching vibration of the hydroxyl group present in the starch molecule. The carbon chain present in the starch molecule that generates the bands observed in 2943 cm^{-1} is due to the C–H stretching vibration; the bands at 1647, 1458, and 1353 cm^{-1} are related to the C–OH bending; and the bands at 1147, 1076, and 980 cm^{-1} correspond to the C–O stretching vibration [45].

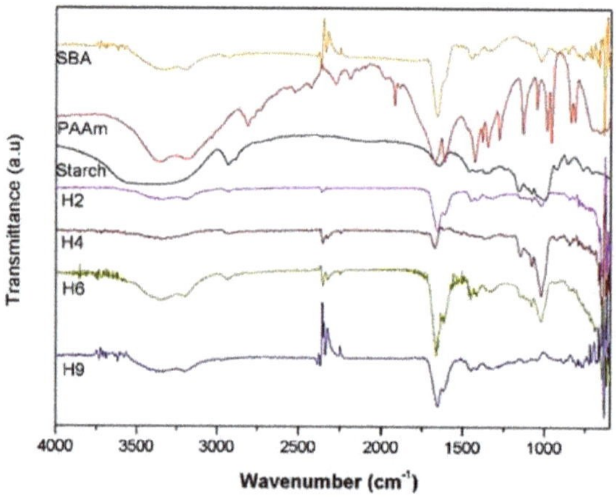

Figure 7. The attenuated total reflectance Fourier transform infrared spectra of sugarcane bagasse ash (SCBA), crosslinked polyacrylamide (PAAm), starch, and the hydrolysed H2, H4, H6, and H9 hydrogels.

It can be noted that the spectra of the hydrolysed hydrogels H2, H4, H6, and H9 are similar and have bands characteristic of both AAm and starch. In the spectrum of the hydrogels, a broad band can be noted at 3500–3100 cm^{-1} that is related to the overlap of the stretching band of the starch hydroxyl group (-OH) with the stretching band of the N-H bond of the amide present in the polyacrylamide, formed by the polymerisation of the acrylamide monomer. In addition, the band around 1700 cm^{-1} corresponds to carboxyl (C=O) also present in acrylamide. The hydrogels H2 and H4 showed a decrease in the band corresponding to carboxyl, possibly due to the hydrolysis reaction, since, in this process, the amide groups (-$CONH_2$) are converted to carboxylate groups (-COO^-Na^+). It can also be noted that the shift of the N-H vibration band of the amide at 1605 cm^{-1} to 1461 cm^{-1} in the hydrogels indicates that crosslinking reactions occurred [24].

The SCBA was obtained through the calcination of SBA at 600 °C. This material had a higher level of pozzolanic activity [23]. Calcination at 600 °C eliminates, by burning, almost all the carbon present in the sample structure [23,24]. The X-ray diffractogram presented in Figure 8 showed an amorphous halo originating from the silica between 2θ of 10° and 50°. Moreover, there are no crystallinity peaks. Therefore, its reactivity is better, and, consequently, it has better applicability to cementitious media [46].

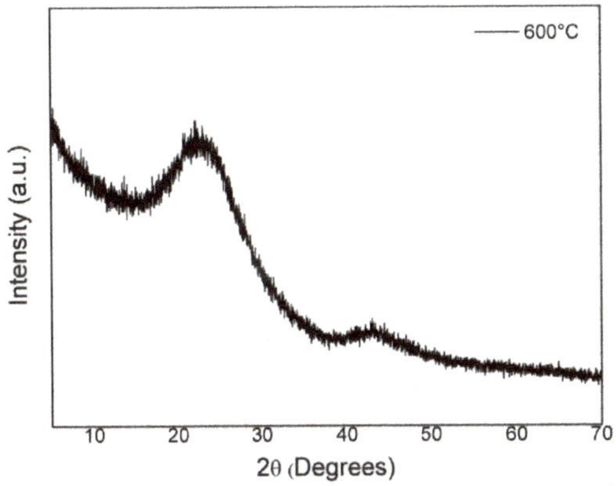

Figure 8. The X-ray diffraction profile of sugarcane bagasse ash obtained by calcination at 600 °C for 2 h.

3.6. FTIR, XRD, and TGA of the Cement Pastes

Figure 9 shows the FTIR spectra of the aged cement pastes. After ageing for 7 and 28 days, there were no changes in the absorption bands in relation to the reference cement paste. This result indicates that the addition of hydrogel into the concrete matrix did not significantly change the composition of the cement pastes. The bands observed at 3629, 3638, and 3658 cm^{-1} correspond to the hydroxyl group (-OH) present in $Ca(OH)_2$; this formation results from the hydration reactions of the compounds C_3S, C_2S, and C-S-H [45]. The bands observed at 1452 and 732 cm^{-1} refer to the -CO bond of the carbonate groups. The band at 991 cm^{-1} is attributed to the asymmetric stretching of the Si-O vibrations of the SiO_4 tetrahedron of the silicate phase present on C_3S and C_2S. In all samples, the vibrations at 713 and 870 cm^{-1} indicate the presence of calcite ($CaCO_3$) [47].

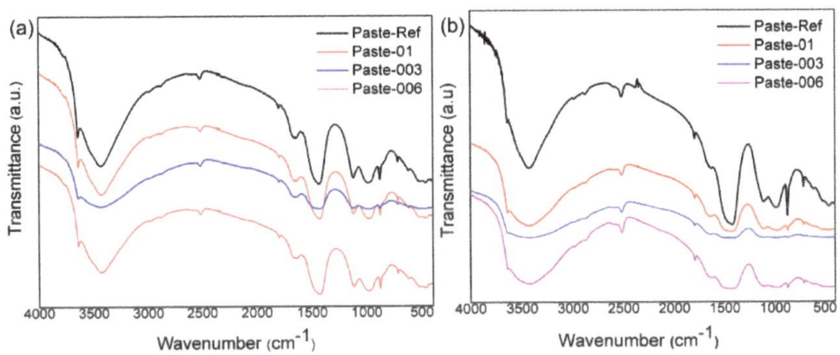

Figure 9. The Fourier transform infrared spectra of cement pastes aged for 7 days (**a**) and 28 days (**b**).

The hydrolysed H2 hydrogel was used to produce the cement pastes due to its degree of swelling from the water absorption and desorption tests (see Table 2 and Figures 2 and 3). Figure 10 shows the XRD profiles obtained in cement pastes after ageing for 7 and 28 days from the reference paste (Paste-ref) and pastes with 0.06%, 0.03%, and 0.1% hydrogel by weight. The pastes with hydrogel and the reference paste (without gel addition) present similar X-ray diffractograms, indicating that the addition of hydrogel did not modify the profiles of the cement paste matrix during curing for 7 and 28 days.

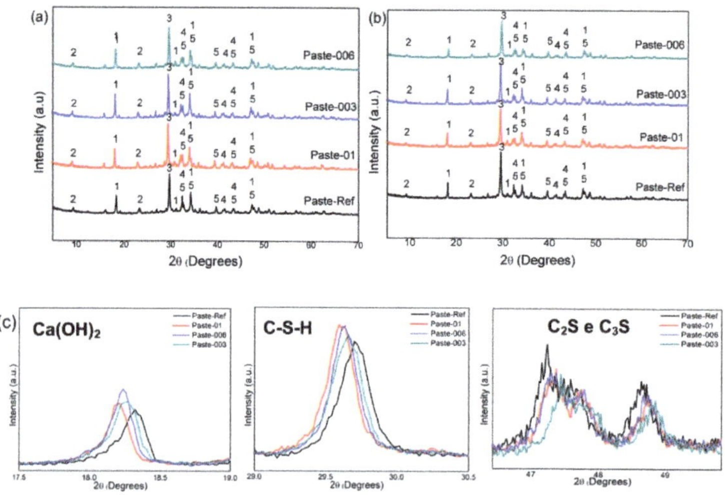

Figure 10. Comparison of the X-ray diffractograms of the cement pastes after ageing for (**a**) 7 and (**b**) 28 days and (**c**) broadening of the $(CaOH)_2$, C-S-H, and C_2S and C_3S peaks.

The X-ray diffractograms were analysed to identify the anhydrous and hydrated crystalline phases in the pastes: (1) portlandite ($Ca(OH)_2$), (2) ettringite, (3) C-S-H, (4) dicalcium silicate (C_2S), and (5) tricalcium silicate (C_3S). In general, portlandite was most detectable, with 2θ peaks of 18°, 31°, 34°, and 46° after 7 and 28 days of curing. Portland cement basically consists of C_2S and C_3S. During the hydration reactions, C_2S and C_3S are transformed into $Ca(OH)_2$ and C-S-H. In other words, the presence (or absence) of C_2S, C_3S, $Ca(OH)_2$, and C-S-H can be noted in the X-ray diffractogram patterns, so it may be indicative of the formation (or not) of such compounds during ageing. The addition of hydrogel to the cement matrix can delay the hydration of the concrete, as it adheres to the particles, thus preventing heat conduction between the cement particles [48,49].

Figure 10c shows that, for all cement pastes containing hydrogel, the peak at 2θ = 18° is more intense than the reference paste for both curing ages studied, i.e., there was greater production of $Ca(OH)_2$, thus indicating better hydration of the medium. Furthermore, it can also be noted that the peak at 2θ = 29° (referring to C-S-H) is also more intense for the pastes containing hydrogel, indicating more formation of C-S-H. It is also possible to note that the peaks referring to C_2S and C_3S (2θ = 39° and 41°) are more intense for the reference paste (without the addition of hydrogel), indicating that this sample contains a greater presence of non-hydrated products, while the samples with hydrogel have smaller peaks of these compounds, indicating that hydration was more efficient [30].

Figure 11 shows the thermogravimetric (TG)/first-derivative thermogravimetric (DTG) curves of the hydrated cement pastes, after ageing for 28 days. The TG/DTG profiles demonstrate the reactions that occurred in the hydrated cement pastes when subjected to a continuous increase in temperature. The TG/DTG curves of the pastes present a peak close to 100 °C; this peak is associated with the release of water and dehydration of C-S-H and ettringite. Due to this overlap, it is not possible to calculate the C-S-H dose accurately. However, another peak around 440 °C corresponds to the dehydration of $Ca(OH)_2$ [6,47,49]. At this temperature, it can be noted that the cement paste without hydrogel (Figure 11a) presented a mass loss of 10%, while those with the addition of hydrogel presented a mass loss of 15%, thus indicating a greater presence of $Ca(OH)_2$ [30].

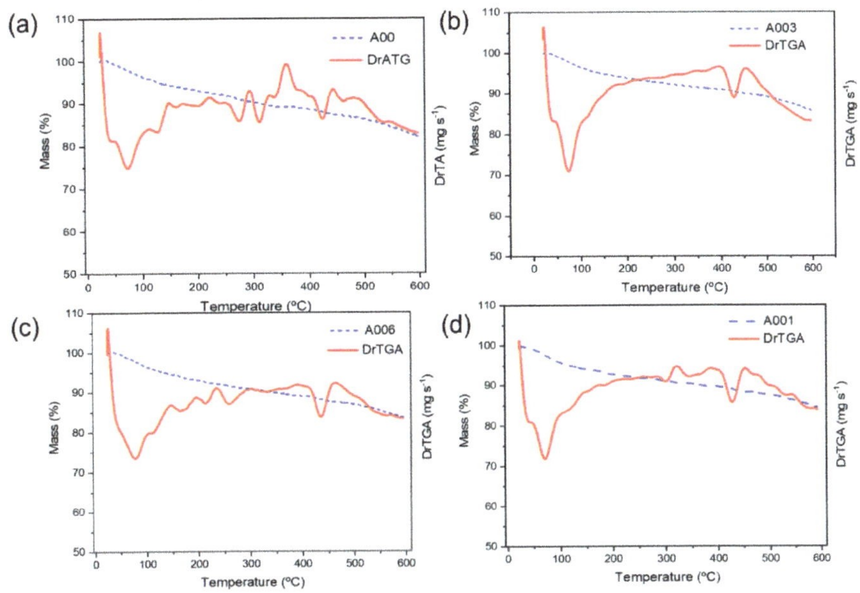

Figure 11. Thermogravimetric (TG)/derivative thermogravimetric (DTG) analysis of Portland cement pastes aged for 28 days: (**a**) reference paste, (**b**) with 0.1% hydrogel by weight, (**c**) with 0.03% hydrogel by weight, and (**d**) with 0.06% hydrogel by weight.

3.7. Compressive Strength Tests of the Mortars

Figure 12 shows the compressive test results for the mortars containing the hydrolysed H2 hydrogel (0.1%, 0.03%, and 0.06% by weight: Mort-01, Mort-003, and Mort-006, respectively). After ageing for 7 days, only Mort-003 presented an increase in resistance relative to the reference mortar (Mort-Ref); the resistance of Mort-006 and Mort-001 decreased relative to Mort-Ref. After ageing for 28 days, similar behaviour was exhibited: an increase for Mort-003 and a reduction for Mort-006 and Mort-01 relative to Mort-Ref.

The addition of 0.03% hydrogel proved to be effective in increasing the mechanical resistance of the mortar. This may be related to the controlled release of water in the cementitious matrix, favouring the curing process [47,50]. This increase can be attributed to the presence of swollen hydrogel in the formulation, because the added hydrolysed H2 hydrogel was swollen at equilibrium. So, in the preparation of the mortar, no additional water was added to the specimens, differently from the cement pastes prepared without hydrogel. The hydrogel released water during ageing, favouring effective curing. The addition of 0.1% and 0.06% hydrogel by weight was unfavourable in terms of resistance. After ageing for 28 days, the discrepancy in the resistance was considerable. The reduction in the strength of Mort-01 and Mort-006 due to the addition of hydrogel can be caused by the formation of voids in the mortar microstructure, precisely because, as water is released by the swollen hydrogels in the cement matrix, there is a reduction in the polymer, causing the formation of voids in the microstructure. The differences in the compressive test results are related to different factors, such as climate and the moulding of the test specimens [51–53]. The compressive test results showed that the developed hydrogels have the capacity to act as an internal curing agent, providing better hydration for the cementitious medium and, consequently, greater mechanical resistance. Furthermore, only a small amount of the hydrolysed H2 hydrogel (0.03% by weight) is necessary to improve the resistance of the material, while higher levels can reduce resistance.

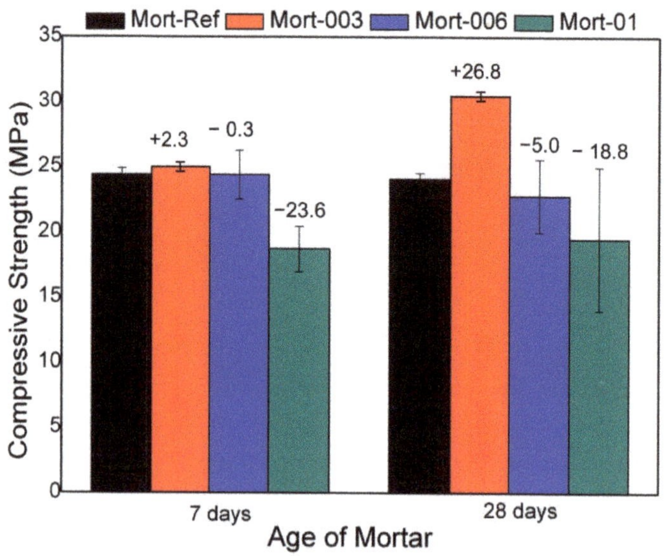

Figure 12. Compressive strength test for the mortars aged for 7 and 28 days. Mort-ref, reference mortar; Mort-003, mortar with 0.03% hydrogel by weight; Mort-006, mortar containing 0.06% hydrogel by weight; and Mort-01, mortar containing 0.1% hydrogel by weight.

The fact that Mort-01 and Mort-006 presented lower resistance compared with Mort-Ref does not mean that the hydrolysed H2 hydrogel cannot act as an internal curing agent. Instead, it can be said that high levels of hydrogel end up releasing a greater amount of water than is necessary to act in the hydration process. This excess water harms the resistance of the material, because a cementitious medium must have a proper balance of its constituent materials [54,55].

In the literature, there are no discussions about the interaction between the hydrogel and the cementitious medium. However, what can occur is the interaction between the polymer matrix and the ions formed in the cement hydration process. As has already been discussed, the presence of ions reduces the absorption of water by the hydrogel. For this reason, in our research, the hydrogels were already swollen when added, so that they could absorb and release the greatest possible amount of water, without being influenced by the environment in which they were inserted.

In addition to hydrogels, another material that is widely studied for use as an internal curing agent is prewetted lightweight aggregates (LWAs), which, like hydrogels, improve relative humidity in the environment by releasing water [56]. However, despite being effective in reducing autogenous shrinkage, the use of LWAs caused a reduction in the mechanical strength of the material, due to the formation of large pores and the increase in the porosity of the material [56–58]. Furthermore, the use of LWAs implies the reduction in fine and coarse aggregates [56,57], which can also be a factor responsible for the reduction of mechanical resistance. Thus, the results obtained can indicate that hydrogels have an advantage over LWAs, as they are capable of reducing autogenous shrinkage and increasing mechanical resistance without the need to reduce another component of the medium.

The water contained in the inner part of the hydrogel is more difficult to be lost through evaporation, as it interacts with the hydrophilic groups present in the polymer matrix, when compared with the free water in the cement pores. In addition, the water absorbed by the hydrogel will be desorbed directly into the cementitious medium, since the hydrogel will be surrounded by the mass formed by the cement hydration process, thus creating another barrier to water loss through evaporation.

The compressive test data were subjected to analysis of variance (ANOVA) to evaluate the statistical significance of the main factors: age (7 and 28 days) and the hydrolysed hydrogel content (0.03%, 0.06%, and 0.01% by weight). Table 3 shows the age, hydrolysed H2 hydrogel content as well as the age × hydrolysed hydrogel content interaction. The age, hydrolysed hydrogel content, and age × hydrogel content interaction were significant. Hence, the amount of hydrolysed hydrogel in the cementitious material (e.g., mortar in this study) and the curing time are highly relevant to the mechanical properties of the cementitious material.

Table 3. Analysis of variance of the compressive strength test data of the mortars, considering the main factors age and the hydrolysed hydrogel content.

Effect	SQ *	DF **	MQ ***	F	p-Value	Result
Age	2.83×10^{14}	11	2.56×10^{13}	3.8	2.3×10^{-3}	Significant
Hydrogel	2.28×10^{14}	6	7.60×10^{13}	11.2	4.0×10^{-4}	Significant
Residue	1.10×10^{14}	-	6.77×10^{12}	-	-	-

* SQ, quadratic sum; ** DF, degrees of freedom; *** MQ, mean of squares.

4. Conclusions

The synthesis of hydrogels from crosslinked PAAm, starch, and SCBA was successful; however, the hydrogels presented low Q values (between 10–15 g/g). After the hydrolysis process, an increase of more than 10 times in the Q value (110–200 g/g) was observed, thus producing superabsorbent hydrogels. SEM images showed that the addition of SCBA did not influence the formation of pores, and SCBA particles were not deposited on the surface of the hydrogel.

FTIR and XRD analyses of the cement pastes indicate that the addition of the hydrogel to the cementitious medium does not alter the composition of the cement pastes, showing that the hydrogel can be added to the cementitious medium without damaging the main characteristics of product. When analysing the mechanical properties, it can be noted that the addition of the 0.03% content was able to increase the mechanical strength in the curing period of 7 and 28 days, indicating that the hydrogel acted as an internal curing agent, improving hydration and, thus, improving the mechanical strength. However, when contents higher than 0.03% (0.06% and 0.1%) were added, a reduction in mechanical strength occurred in up to 26.8% at 28 days of curing, as compared with reference (without hydrogel). This does not mean that the hydrogel did not act as an internal curing agent and is related to the formation of voids after the release of water by the hydrogel, which negatively influences the strength of the material.

In addition to what was reported above, to the best of our knowledge, this is the first time that a hydrogel containing crosslinked PAAm, starch, and SCBA has been used as a curing agent for mortars and cement pastes. In addition to the technological use of hydrogel as a curing agent, this approach is environmentally friendly, because it uses a natural product (starch) and SCBA, a byproduct of the sugarcane industry. This research paves the way for the development of new research using hydrogels containing crosslinked PAAm, starch, and SCBA with application in the cementitious medium (using the hydrogels that have not yet been applied in the cementitious medium) and for new applications that require materials that can absorb and release large amounts of water.

Author Contributions: A.E.N.P.: data curation, investigation, methodology, software, validation, writing—original draft; E.A.d.A.: data curation, investigation, methodology, software, validation, writing—original draft; F.R.K.: investigation, methodology, software, validation; E.C.d.S.-F.: data curation, software, validation, co-supervision, writing—original draft; E.C.M.: conceptualization, funding acquisition, supervision, writing—original draft, writing—review and editing. All authors have read and agreed to the published version of the manuscript.

Funding: This research was funded by the Conselho Nacional de Desenvolvimento Científico e Tecnológico (CNPq) [grant #406973/2022-9, #307429/2018-0 and #408767/2021-9].

Institutional Review Board Statement: Not applicable.

Informed Consent Statement: Not applicable.

Data Availability Statement: The original contributions presented in the study are included in the article, further inquiries can be directed to the corresponding author.

Conflicts of Interest: The authors declare no conflict of interest. The funder was not involved in the study design, collection, analysis, interpretation of data, the writing of this article or the decision to submit it for publication.

References

1. Kolawole, J.T.; Babafemi, A.J.; Fanijo, E.; Paul, S.C.; Combrinck, R. State-of-the-art review on the use of sugarcane bagasse ash in cementitious materials. *Cem. Concr. Compos.* **2021**, *118*, 103975. [CrossRef]
2. Kasprzhitskii, A.S.; Kruglikov, A.A. Molecular Insights into Adhesion at Interface of Geopolymer Binder and Cement Mortar. *Int. J. Molec. Sci.* **2024**, *25*, 8374. [CrossRef]
3. Calesco, M.A.F.; Watanuki Filho, A.; Moura, M.R.; Aouada, F.A. Effects of addition of hybrid nanocomposite based on polysaccharide hydrogel and nanoclay on the fresh and hardened properties of cementitious mortars. *Cerâmica* **2022**, *68*, 97–107. [CrossRef]
4. Souza, B.C.; Paes, C.N.S.; Barboza, L.S. Study of aggregate graft recycled of concrete as an internal healing agent. *Rev. Mater.* **2020**, *25*, 12869.
5. Rathinam, K.; Murali, A.; Pershiya, D.; Kaviarasu, S. Properties of concrete incorporated with self -healing and internal curing agents. *J. Teknol.* **2024**, *86*, 145–154. [CrossRef]
6. Manzano, M.A.R.; Silva, E.F.; Lopes, A.N.M.; Tolêdo Filho, R.D. Actuating mechanism of Superabsorbent Polymers as internal curing agents to mitigate autogenous shrinkage in High Strength Concrete (HSC)—State of the art. *Rev. Mater.* **2021**, *26*, 12956.
7. Neville, A.A. *Concrete Properties*, 5th ed.; Bookman: London, UK, 2016; pp. 1–800.
8. Chajec, A.; Chowaniec, A.; Królicka, A.; Sadowski, L.; Zak, A.; Piechowka-Mielnik, M.; Savija, B. Engineering of green cementitious composites modified with siliceous fly ash: Understanding the importance of curing conditions. *Constr. Build. Mater.* **2021**, *313*, 125209. [CrossRef]
9. Jalal, A.; Kiranb, R. Quantifying the water donation potential of commercial and corn starch hydrogels in a cementitious matrix. *J. Mater. Res. Technol.* **2023**, *24*, 4336–4352. [CrossRef]
10. Gwon, S.; Seifu, M.N.; Shin, M.; Park, S. Water migration mechanism during internal curing of cement composites with cellulose microfiber. *Constr. Build Mater.* **2024**, *411*, 134173. [CrossRef]
11. Wang, H.; Habibi, M.; Marzouki, R.; Majdi, A.; Shariati, M.; Denic, N.; Zakic, A.; Khorami, M.; Khadimallah, A.A.; Ebid, A.A.K. Improving the Self-Healing of Cementitious Materials with a Hydrogel System. *Gels* **2022**, *8*, 278. [CrossRef]
12. Hussain, S.; Salman, M.; Youngblodd, J.P.; Farooq, U.; Yasmeen, S.; Al-Ahmary, K.M.; Ahmed, M. Enhanced adsorption of Congo red dye by CS/PEG/ZnO composite hydrogel: Synthesis, characterization, and performance evaluation. *J. Mol. Liq.* **2024**, *411*, 125704. [CrossRef]
13. Gyles, D.A.; Castro, L.D.; Silva, J.O.C., Jr.; Ribeiro-Costa, R.M. A review of the designs and prominent biomedical advances of natural and synthetic hydrogel formulations. *Eur. Polym. J.* **2017**, *88*, 373–392. [CrossRef]
14. Zhang, X.; Yao, Y.; Wu, Y.; Wang, X.; Feng, P.; Zhang, J.; Hu, W.; Shang, E. Enhancement and mechanism of mechanical properties and functionalities of polyacrylamide/polyacrylic acid hydrogels by 1D and 2D nanocarbon. *J. Colloid Interface Sci.* **2024**, *679*, 79–89. [CrossRef] [PubMed]
15. Shahid, B.; Hina, M.; Iqbal, J.; Rajpar, A.H.; Mujtaba, M.A.; Alghamdi, N.A.; Wageh, S.; Ramesh, K.; Ramesh, S. Fundamental Concepts of Hydrogels: Synthesis, Properties, and Their Applications. *Polymers* **2020**, *12*, 2702. [CrossRef] [PubMed]
16. Tetlow, I.J.; Bertoft, G. A Review of Starch Biosynthesis in Relation to the Building Block-Backbone Model. *Int. J. Mol. Sci.* **2020**, *21*, 7011. [CrossRef] [PubMed]
17. Dong, Y.; Ghasemzadeh, M.; Khirsandi, Z.; Sheibani, R.; Nasrollahzadeh, M. Starch-based hydrogels for environmental applications: A review. *Int. J. Biol. Macromol.* **2024**, *269*, 131956. [CrossRef] [PubMed]
18. Sohail, M.; Mudassir; Minhas, M.U.; Khan, S.; Hussain, Z.; de Matas, M.; Shah, S.A.; Khan, S.; Kousar, M.; Ullah, K. Natural and synthetic polymer-based smart biomaterials for management of ulcerative colitis: A review of recent developments and future prospects. *Drug Deliv. Transl. Res.* **2018**, *9*, 595–614.
19. Abbasi, M.; Sohail, M.; Minhas, M.U.; Khan, S.; Hussain, Z.; Mahmood, A.; Shah, A.S.; Kousar, M. Novel biodegradable pH-sensitive hydrogels: An efficient controlled release system to manage ulcerative colitis. *Int. J. Biol. Macromol.* **2019**, *136*, 83–96. [CrossRef]
20. Qamruzzaman, M.; Ahmed, F.; Mondal, M.I.H. An Overview on Starch-Based Sustainable Hydrogels: Potential Applications and Aspects. *J. Polym. Environ.* **2022**, *30*, 19–50. [CrossRef]
21. El Sayed, M.M. Production of Polymer Hydrogel Composites and Their Applications. *J. Polym. Environ.* **2023**, *31*, 2855–2879. [CrossRef]

22. Bortolin, A.; Aouada, F.A.; Longo, E.; Mattoso, L.H.C. Investigation of water absorption process in polysaccharide hydrogels: Effect of ionic charge, presence of salt, monomer and polysaccharide concentration. *Polímeros* **2012**, *22*, 311–317. [CrossRef]
23. Athira, G.; Bahurudeen, A. Rheological properties of cement paste blended with sugarcane bagasse ash and rice straw ash. *Constr. Build Mater.* **2022**, *332*, 127377. [CrossRef]
24. Khalil, M.J.; Aslam, M.; Ahmad, S. Utilization of sugarcane bagasse ash as cement replacement for the production of sustainable concrete–A review. *Constr. Build Mater.* **2021**, *270*, 121371. [CrossRef]
25. Associação Brasileira de Normas Técnicas. *NBR 7214: Areia Normal para Ensaio de Cimento—Especificação;* ABNT: Rio de Janeiro, Brazil, 2015.
26. Rodrigues, R.H.S.; Almeida, E.A.; Kruger, F.R.; Silva Filho, E.C.; Muniz, E.C. Hydrogels Based on Polyacrylamide, Pectin and Containing Rice Husk Ash: Preparation, Characterization and Application in Formulation of Cementicious Materials. *Materials* **2024**, *17*, 5746. [CrossRef]
27. Sousa, H.R.; Lima, I.S.; Neris, L.M.L.; Silva, A.S.; Santos Nascimento, A.M.S.; Araújo, F.P.; Ratke, R.F.; Silva, D.A.; Osajima, J.A.; Bezerra, L.R.; et al. Superabsorbent Hydrogels Based to Polyacrylamide/Cashew Tree Gum for the Controlled Release of Water and Plant Nutrients. Molecules. *Molecules* **2021**, *26*, 2680. [CrossRef]
28. Associação Brasileira de Normas Técnicas. *NBR 7215: Cimento Portland–Resistência à Compressão do Cimento Portland;* ABNT: Rio de Janeiro, Brazil, 1997.
29. Agostinho, L.B.; Pereira, A.C.; da Silva, E.F.; Toledo Filho, R.D. Rheological study of Portland cement pastes modified with superabsorbent polymer and nanosilica. *J. Build. Eng.* **2021**, *34*, 102024. [CrossRef]
30. Esteves, L.P.; Lukošiūtė, I.; Čėsnienė, J. Hydration of cement with superabsorbent polymers. *J. Therm. Anal. Calorim.* **2014**, *118*, 1385–1393. [CrossRef]
31. Agostinho, L.B.; Borges, J.G.; Silva, E.F.; Cupertino, D.V.M.R. A calorimetry analysis of Portland cement pastes containing superabsorbent polymer (SAP) and nanosilica (NS). *Rev. Mater.* **2020**, *25*, 12860.
32. Kim, S.; Iyer, G.; Nadarajah, A.; Frantz, J.M.; Spongberg, A.L. Polyacrylamide Hydrogel Properties for Horticultural Applications. *Int. J. Polym. Anal. Charact.* **2010**, *15*, 307–318. [CrossRef]
33. Krafcik, M.J.; Macke, D.N.; Erk, K.A. Improved Concrete Materials with Hydrogel-Based Internal Curing Agents. *Gels* **2017**, *3*, 46. [CrossRef]
34. Tanwar, A.; Date, P.; Ottoor, D. ZnO NPs incorporated gelatin grafted polyacrylamide hydrogel nanocomposite for controlled release of ciprofloxacin. *J. Colloid Interface Sci.* **2021**, *42*, 100413. [CrossRef]
35. Zhao, Q.; Sun, J.; Lin, Y.; Zhou, Q. Study of the properties of hydrolyzed polyacrylamide hydrogels with various pore structures and rapid pH-sensitivities. *React. Funct. Polym.* **2010**, *70*, 602–609. [CrossRef]
36. Jayaramudu, T.; Ko, H.-U.; Kim, H.C.; Kim, J.W.; Kim, J. Swelling Behavior of Polyacrylamide–Cellulose Nanocrystal Hydrogels: Swelling Kinetics, Temperature, and pH Effects. *Materials* **2019**, *12*, 2080. [CrossRef] [PubMed]
37. Bennour, S.; Louzri, F. Study of Swelling Properties and Thermal Behavior of Poly (N, N-Dimethylacrylamide- co -Maleic Acid) Based Hydrogels. *Adv. Chem.* **2014**, *2014*, 147398. [CrossRef]
38. Erizal, E. Synthesis of poly(acrylamide-co-acrylic acid)-starch based superabsorbent hydrogels by gamma radiation: Study its swelling behavior. *Indones. J. Chem.* **2012**, *12*, 113–118. [CrossRef]
39. Mahon, R.; Balogun, Y.; Oluyemi, G.; Njuguna, J. Swelling performance of sodium polyacrylate and poly(acrylamide-co-acrylic acid) potassium salt. *SN Appl. Sci.* **2020**, *2*, 117. [CrossRef]
40. Garcia, J.A.F.; Moura, M.R.; Aouada, F.A. Effect of ph, ionic concentration and species on the absorption of water by hydrogel bionanocomposits constituted from CMC/PAAm/LAPONITE RDS. *Quím. Nova* **2019**, *42*, 831–837.
41. Awasthi, S.; Gaur, J.K.; Bobji, M.S.; Srivastava, C. Nanoparticle-reinforced polyacrylamide hydrogel composites for clinical applications: A review. *J. Mater. Sci.* **2022**, *57*, 8041–8063. [CrossRef]
42. Tavlieva, M.P.; Genieva, S.D.; Georgieva, V.G.; Vlaec, L.T. Thermodynamics and kinetics of the removal of manganese (II) ions from aqueous solutions by white rice husk ash. *J. Mol. Liq.* **2015**, *211*, 938–947. [CrossRef]
43. Kernosenko, L.; Samchenko, K.; Goncharuk, O.; Pasmurtseva, N.; Poltoratska, T.; Siryk, O.; Dziuba, O.; Mironov, O.; Szewczuk-Karpisz, K. Polyacrylamide Hydrogel Enriched with Amber for In Vitro Plant Rooting. *Plants* **2023**, *12*, 1196. [CrossRef]
44. Mu, X.; Zhou, J.; Wang, P.; Chen, H.; Yang, T.; Chen, S.; Miao, L.; Mori, T. A robust starch–polyacrylamide hydrogel with scavenging energy harvesting capacity for efficient solar thermoelectricity–freshwater cogeneration. *Energy Environ. Sci.* **2022**, *15*, 3388–3399. [CrossRef]
45. Torres-Figueroa, A.V.; Pérez-Martínez, C.J.; del Castillo-Castro, T.; Bolado-Martínez, E.; Corella-Madueño, M.A.G.; García-Alegría, A.M.; Laran-Ceniceros, T.E.; Armenta-Villegas, L. Composite Hydrogel of Poly(acrylamide) and Starch as Potential System for Controlled Release of Amoxicillin and Inhibition of Bacterial Growth. *J. Chem.* **2020**, *2020*, 5860487. [CrossRef]
46. Bahurudeen, A.; Santhanam, M. Influence of different processing methods on the pozzolanic performance of sugarcane bagasse ash. *Cem. Concr. Compos.* **2015**, *56*, 32–45. [CrossRef]
47. He, P.; Shi, C.; Tu, Z.; Zhang, J. Effect of further water curing on compressive strength and microstructure of CO_2-cured concrete. *Cem. Concr. Compos.* **2016**, *72*, 80–88. [CrossRef]
48. Gomes, C.M.; Cheung, N.; Gomes, G.M.; Sousa, A.K.; Peruzzi, A.P. Improvement of water resistance in magnesia cements with renewable source silica. *Constr. Build Mater.* **2021**, *272*, 121650. [CrossRef]

49. Secchi, M.; Neves Junior, A. Adição de cinzas do bagaço da cana-de-açúcar em compósitos cimentícios reforçados com fibras de sisal. *E&S Eng. Sci.* **2019**, *8*, 2–14.
50. Almeida, E.A. Synthesis and Characterization of Polyacrylamide Hydrogels with Rice Husk Ash for Application in Cementitious Materials as an Internal Curing Agent. Master's Thesis, Department of Chemistry, State University of Maringa, Maringa, Brazil, 2022.
51. Snoeck, D.; Tittelboom, V.K.; Steuperaert, S.; Dubruel, P.; De Belie, N. self-healing cementitious materials by the combination of microfibres and superabsorbent polymers. *J. Intell. Mater. Syst. Struct.* **2014**, *25*, 13–24. [CrossRef]
52. Ma, X.; Liu, J.; Wu, Z.; Shi, C. Effects of SAP on the properties and pore structure of high-performance cement-based materials. *Constr. Build. Mater.* **2017**, *131*, 476–484. [CrossRef]
53. Schröfl, C.; Mechtcherine, V.; Gorges, M. Relation between the molecular structure and the efficiency of superabsorbent polymers (SAP) as concrete admixture to mitigate autogenous shrinkage. *Cem. Conc. Res.* **2012**, *42*, 865–873. [CrossRef]
54. Kang, S.H.; Hong, S.G.; Moon, J. Importance of drying to control internal curing effects on field casting ultra-high-performance concrete. *Cem. Conc. Res.* **2018**, *108*, 20–30. [CrossRef]
55. Cui, L.; Wang, H. Research on the Mechanical Strengths and the Following Corrosion Resistance of Inner Steel Bars of RPC with Rice Husk Ash and Waste Fly Ash. *Coatings* **2021**, *11*, 1480. [CrossRef]
56. Wen, C.; Shen, D.; Feng, Z.; Liu, C.; Deng, S. Relationship between internal relative humidity and autogenous shrinkage of early-age concrete containing prewetted lightweight aggregates. *Struct. Concr.* **2022**, *24*, 4110–4125. [CrossRef]
57. Choi, S. Internal relative humidity and drying shrinkage of hardening concrete containing lightweight and normal-weight coarse aggregates: A comparative experimental study and modeling. *Constr. Build Mater.* **2017**, *148*, 288–296. [CrossRef]
58. Bentur, A.; Igarashi, S.I.; Kovler, K. Prevention of autogenous shrinkage in high-strength concrete by internal curing using wet lightweight aggregates. *Cem. Conc. Res.* **2001**, *31*, 1587–1591. [CrossRef]

materials

MDPI

Article

Degradation of Poly(ethylene terephthalate) Catalyzed by Nonmetallic Dibasic Ionic Liquids under UV Radiation

Ruiqi Zhang [1,2], Xu Zheng [3], Xiujie Cheng [1,2], Junli Xu [1,2], Yi Li [1,2], Qing Zhou [1,2], Jiayu Xin [1,2], Dongxia Yan [1,2] and Xingmei Lu [1,2,*]

1 Beijing Key Laboratory of Ionic Liquids Clean Process, CAS Key Laboratory of Green Process and Engineering, State Key Laboratory of Multiphase Complex Systems, Institute of Process Engineering, Chinese Academy of Sciences, Beijing 100190, China; rqzhang@ipe.ac.cn (R.Z.); jyxin@ipe.ac.cn (J.X.)
2 School of Chemical Engineering, University of Chinese Academy of Sciences, Beijing 100049, China
3 College of Chemistry, Liaoning University, Shenyang 110036, China
* Correspondence: xmlu@ipe.ac.cn

Abstract: Nonmetallic ionic liquids (ILs) exhibit unique advantages in catalyzing poly (ethylene terephthalate) (PET) glycolysis, but usually require longer reaction times. We found that exposure to UV radiation can accelerate the glycolysis reaction and significantly reduce the reaction time. In this work, we synthesized five nonmetallic dibasic ILs, and their glycolysis catalytic activity was investigated. 1,8-diazabicyclo [5,4,0] undec-7-ene imidazole ([HDBU]Im) exhibited better catalytic performance. Meanwhile, UV radiation is used as a reinforcement method to improve the PET glycolysis efficiency. Under optimal conditions (5 g PET, 20 g ethylene glycol (EG), 0.25 g [HDBU]Im, 10,000 $\mu W \cdot cm^{-2}$ UV radiation reacted for 90 min at 185 °C), the PET conversion and BHET yield were 100% and 88.9%, respectively. Based on the UV-visible spectrum, it was found that UV radiation can activate the C=O in PET. Hence, the incorporation of UV radiation can considerably diminish the activation energy of the reaction, shortening the reaction time of PET degradation. Finally, a possible reaction mechanism of [HDBU]Im-catalyzed PET glycolysis under UV radiation was proposed.

Keywords: PET glycolysis; nonmetallic dibasic ionic liquids; UV radiation; catalyst

1. Introduction

PET has an important position in the field of plastic applications since it is non-toxic, odorless, and tasteless [1]. These unique qualities allow PET to be employed in diverse textiles and soft drink bottles [2,3]. However, it poses a serious threat to both the environment and ecosystems due to the accumulation of large quantities of PET waste [4]. Furthermore, while PET itself is considered nontoxic, certain additives and colorants present in PET products have the potential to accumulate in the body and lead to toxicity in organisms [5]. And under 100% humidity conditions, the life expectancy of a PET bottle is approximately 93 years [6]. Therefore, achieving complete degradation of PET in natural environments is an almost impossible challenge. In addition, approximately seven barrels of petroleum resources can be saved for every ton of mixed plastic recycled [7]. Hence, the recycling of waste PET holds significant importance for the environment, resources, and economy.

Chemical recycling of waste PET has garnered increasing attention as a means to achieve a closed-loop process and diminish the reliance on fossil energy resources. It facilitates the efficient utilization of waste PET by depolymerizing it into monomers or other chemical compounds [8,9]. Various chemical manners have been studied including glycolysis [10], hydrolysis [11], methanolysis [12], and aminolysis [13]. Among them, glycolysis has attracted great attention because of its mild reaction conditions and because no toxic substances are released during the reaction. PET glycolysis is a process of transesterification with EG at a certain temperature to generate bis (hydroxyethyl) terephthalate (BHET). In the absence of a catalyst, the reaction takes place very slowly [14]. Therefore, various glycolysis catalysts have been developed [15]. Some catalysts, such as ZnO [16], Mn_3O_4 [16], zeolite [17], and SO_4^{2-}/ZnO [18] have been adopted, but they can not dissolve in EG [19]. Thus, the catalytic activity is relatively low due to the complexity of mass transfer. Compared with these solid catalysts, ILs exhibit superior catalytic activity. In 2009, our team first used [Bmim]Cl as a solvent to dissolve and degrade PET [20]. Since then, more and more imidazolium-based ILs have been adopted, such as $[Bmim]_2[CoCl_4]$ [21], $[Amim][ZnCl_3]$ [22], and $[Bmim][ZnCl_3]$ [23]. Similar to metal salts, ILs that contain Zn^{2+} exhibit excellent catalytic efficiency. However, the residual Zn^{2+} in BHET can seriously downgrade the quality of BHET and recycled PET. Therefore, to solve this issue, some nonmetallic ILs catalysts were designed, such as [Ch][OAc] [24] and $[Ch]_3[PO_4]$ [25]. All of them are biocompatible ILs, but it takes more than three hours to fully degrade PET, which leads to high energy consumption and is not conducive to large-scale production. Therefore, it is imperative to improve the reaction rate of nonmetallic IL catalysts. Previous researchers have indicated that alkaline ILs often exhibit better catalytic activity than acidic ILs in PET glycolysis reactions. The catalytic performance of DBU, which has stronger alkalinity, has been demonstrated [26]. Additionally, imidazole anion ([Im]-) also exhibits alkalinity. Therefore, it is considered desirable to design and synthesize [HDBU]Im as a nonmetallic dibasic catalyst, with the anticipation of showcasing remarkable catalytic efficacy in the PET glycolysis reaction.

On the other hand, researchers have developed various auxiliary promotion methods to further enhance PET glycolysis reactions. Alnaqbi et al. [27] used [Bmim]Br as a catalyst for PET glycolysis under microwaves, which could reduce the glycolysis time from nearly 9 h (conventional heating) to 2 h, dramatically increasing the catalytic reaction rate. Imran et al. [28] utilized supercritical ethylene glycol (450 °C and 15.3 MPa) for PET degradation, resulting in a BHET yield of 93.5% within 30 min. The shortened reaction time is attributed to its solubility and high solvent density. However, the temperature and pressure under supercritical conditions are relatively high. Le et al. [29] developed a co-solvent-assisted PET glycolysis method that can complete PET decomposition by reacting at 153 °C for 2 h using anisole as a co-solvent. However, anisole has a detrimental impact on the environment. As part of the sunlight spectrum, ultraviolet (UV) radiation is cheaper, more accessible, and more environmentally friendly than other auxiliary promotion methods. UV radiation has been identified as the primary factor for plastic degradation in the natural environment [30]. MacLeod et al. [31] reported that under marine conditions, the photo-induced oxidation of PET is likely to occur, which leads to a reduction in molecular weight. In this process, the function of UV radiation is to induce the cleavage of the carbon–carbon backbone, leading to chain scission. Lee et al. [32] identified the photo-degradation products of PET films including esters, peresters, and benzoic acids. More et al. [33] applied UV radiation in PET aminolysis. It can be found that the speed of the degradation process was enhanced, which was due to the UV radiation attacking the ester linkage of the PET [33]. Therefore, UV radiation can be used as an efficacious method to promote the occurrence of PET degradation. However, the research on IL-catalyzed PET glycolysis under UV radiation-assisted systems is still limited. Motivated by this, UV radiation was introduced into the PET degradation system to enhance the catalytic reaction rate of nonmetallic ILs and the yield of BHET.

In this study, UV radiation was introduced into the IL-catalyzed PET glycolysis system as an auxiliary strengthening method. Meanwhile, five nonmetallic dibasic ILs catalysts were synthesized, and their activity was tested under the same conditions. Among them, [HDBU]Im exhibited the best activity. The optimization of the reaction conditions was achieved through the evaluation of influencing parameters. The reaction kinetics under UV radiation and without UV radiation were studied, and the impact of UV radiation on the reaction mechanism was subsequently examined. Finally, based on the experimental findings a potential degradation mechanism was suggested. Compared with conventional heating methods, UV radiation can shorten the reaction time of nonmetallic IL-catalyzed PET glycolysis and increase the yield of BHET. This approach conserves energy and processing expenses without using hazardous reagents, offering a novel strategy for enhancing the efficiency and sustainability of PET chemical recycling.

2. Materials and Methods

2.1. Materials

Raw PET material was acquired from DuPont (Wilmington, DE, USA). They were smashed to 40–60 mesh. At this particle size, PET has a large surface area, which is conducive to mutual contact between PET, EG, and IL, which can speed up the reaction rate and achieve faster degradation of PET. The molecular weight (M_n) of the PET powder was 4.35×10^4 g·mol^{-1}, determined using GPC analysis. Imidazole (Im), 2-methylimidazole (2-MeIm), 2-ethylimidazole (2-EtIm), 4-methylimidazole (4-MeIm), and 2-ethyl-4-methylimidazole (2-Et-4-MeIm) were purchased from Sinopharm Chemical Reagent Beijing Co., Ltd., Beijing, China.

2.2. Synthesis of ILs

A series of nonmetallic dibasic ILs were synthesized according to reported procedures [34,35]. Taking [HDBU]Im as an example, it was synthesized through the neutralization reaction of DBU and imidazole. A certain amount of imidazole was added dropwise to the anhydrous ethanol solution of DBU. The molar amount of imidazole in the mixture is equimolar to DBU. DBU, imidazole, and anhydrous ethanol were stirred vigorously at 25 °C for 24 h. The anhydrous ethanol was removed using vacuum evaporation, and a pale yellow transparent liquid was obtained. The synthesis steps of 1,8-diazabicyclo [5,4,0] undec-7-ene 2-methylimidazole ([HDBU][2-MeIm]), 1,8-diazabicyclo [5,4,0] undec-7-ene 2-ethylimidazole ([HDBU][2-EtIm]), 1,8-diazabicyclo [5,4,0] undec-7-ene 4-methylimidazole ([HDBU][4-MeIm]), and 1,8-diazabicyclo [5,4,0] undec-7-ene 4-methylimidazole 2-ethyl-4-methylimidazole ([HDBU][2-Et-4-MeIm]) were similar to [HDBU]Im.

2.3. PET Glycolysis under UV Radiation

The reaction was conducted in a dark box under UV radiation. In each experiment, a 50 mL three-neck flask containing a thermometer, a condensation device, and a magnetic stirrer was utilized, and 5.0 g PET powder, 20 g EG, and a definite amount of catalyst were added. They were heated in an oil bath. The temperature was from 160 °C to 195 °C and the reaction time was from 20 min to 150 min. After the completion of degradation, the mixture was cooled to 25 °C, then poured into 500 mL deionized water and stirred vigorously at 70 °C for 1 h to separate the unreacted PET particles through filtration. The unreacted PET was dried at 65 °C for 12 h, and it was weighted to compute PET conversion, which was defined using Equation (1)

$$\text{PET Conversion} = \frac{W_0 - W_1}{W_0} \times 100\% \quad (1)$$

where W_0 represents the initial weight of the PET and W_1 represents the weight of the unreacted PET. The filtrate was subsequently subjected to rotary evaporation in a vacuum at 65 °C, and the concentrated filtrate was cooled at 4 °C for 12 h. Finally, a needle-like BHET monomer was obtained through filtration and drying. The BHET yield is defined using Equation (2):

$$\text{BHET Yield} = \frac{W_{BHET}/M_{BHET}}{W_0/M_{PET}} \quad (2)$$

where W_{BHET} is the weight of BHET and M_{BHET} and M_{PET} are the molecular weight of BHET and the PET repeat unit, respectively.

2.4. Recycling of the EG and Catalyst

After the BHET was extracted through filtration, the remaining EG and catalyst in the filtrate were subjected to rotary evaporation at 65 °C and then stored in a vacuum oven at 70 °C for over 12 h to remove as much water as possible. In the succeeding cycle, a certain amount of the EG was added to ensure that the quantity of the solution was equal to 20 g.

2.5. Characterization

The ^{1}H NMR spectra were analyzed using an AVANCE-III 600 NMR spectrometer (Bruker, Fällanden, Switzerland), operated at 600 MHz in $CDCl_3$ and DMSO-d_6. The FT-IR testing of the catalysts and main product were performed with a Nicolet 380 FT-IR spectrograph (Thermo Nicolet, Waltham, MA, USA). The differentials scanning calorimeter (DSC) STARe system was applied to generate the DSC curve from 25 °C to 200 °C at the heating rate of 10 °C/min under a nitrogen atmosphere. The thermogravimetric analysis (TGA) curve was tested using DTG-60H (Shimadzu, Tokyo, Japan) under the nitrogen atmosphere, and the samples were heated from 25 °C to 600 °C at the heating rate of 10 °C/min. Gel permeation chromatography (GPC) (Agilent, PL-50, Santa Clara, CA, USA) was applied to determine the molecular weights of PET pellets and oligomers. The testing condition was as follows: the oven temperature was 30 °C, the mobile phase was chloroform.

3. Results

3.1. Screening of Catalysts

To investigate the effect of imidazole derivatives anions on catalytic activity, a series of nonmetallic dibasic ILs were prepared. Their structures are shown in Scheme 1. The steric-hindrance effect of anions can significantly affect the catalytic activity. From Table 1, it is evident that [HDBU][Im] exhibits the highest catalytic activity, with a BHET yield of 82.9%. It was also found that when the position of 2-H, 4-H, and 5-H of the imidazole ring are replaced by other functional groups, the catalytic activity can be reduced. For entry 2 and entry 3, the catalytic activity is similar. Therefore, the position of the substituent group on the imidazole ring has relatively tiny influence on catalytic activity. For entry 3 and entry 4, if the volume of the substituent group is bigger, the steric-hindrance effect is noticeable and the hydrogen bonds between the catalyst and the EG will be weakened. Meanwhile, the electron density will be reduced. Thus, the catalytic activity decreases. It can be proved that the 2-H of the imidazole ring plays a crucial role in catalyzing PET glycolysis. 2-H is more active than others, which can leave the imidazole ring and become H^+. The H^+ can attack the carbon of C=O in PET, and then the electrophilicity will be improved. Thus, [HDBU][Im] is the best catalyst, and the structure of the [HDBU][Im] was characterized using FT-IR and ^{1}H NMR (see Supplementary Materials). On the other hand, through a series of characterizations of the main product, it was proved that the main product obtained from the degradation of PET catalyzed by [HDBU][Im] is the monomer BHET (see Supplementary Materials for more detail).

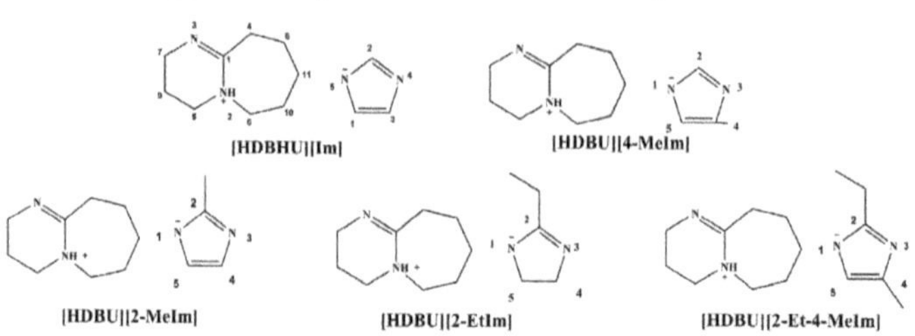

Scheme 1. Molecular formula of DBU-based ionic liquids (atoms were numbered).

Table 1. Different catalysts activities for PET glycolysis [1].

Entry	ILs	PET Conversion (%)	BHET Yield (%)
1	[HDBU][Im]	100	82.9
2	[HDBU][4-MeIm]	100	78.3
3	[HDBU][2-MeIm]	100	77.2
4	[HDBU][2-EtIm]	96.2	74.1
5	[HDBU][2-Et-4-MeIm]	100	70.5

[1] Reaction conditions: 5.0 g PET, 5.0 wt% catalyst, 20.0 g EG reacted for 90 min at 190 °C under 5000 μW·cm^{-2} UV radiation.

3.2. The Effect of UV Radiation

In order to explore the effect of UV radiation on this reaction, the conversion of PET and the yield of BHET were compared with and without UV radiation. The experimental data are shown in Figure 1. UV radiation has been shown to enhance the depolymerization reaction of PET, resulting in a 100% conversion rate after 30 min compared to 120 min without UV radiation. Therefore, UV radiation can significantly accelerate the reaction and shorten the reaction time, and it can also increase the yield of BHET. This may be because UV radiation can degrade the polymer into oligomers effectively. To verify this conjecture, the oligomers generated under UV radiation and without UV were separated and analyzed using GPC (Table 2). UV radiation reduced the M_n of oligomers by approximately 50%. This shows that UV does accelerate polymer degradation, so more oligomers with lower molecular weight were produced. Son et al. [36] found that UV radiation can reduce the activation energy of the reaction system. Therefore, it is speculated that in this glycolysis reaction, UV radiation can also reduce the activation energy, accelerating the reaction rate. The effect of UV radiation on activation energy will be discussed in detail in subsequent sections.

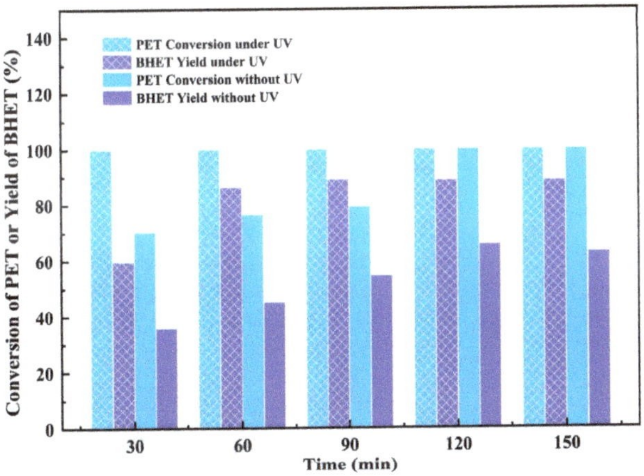

Figure 1. Effect of UV on PET conversion and BHET yield at different reaction time (reaction conditions: 5 g PET, 5 wt% catalyst/PET ratio, 1/4 PET/EG weight ratio at 185 °C).

Table 2. GPC analysis for oligomers.

Condition	M_n (g·mol^{-1})	M_w (g·mol^{-1})	PD
Without UV radiation	2419	2461	1.02
With UV radiation	1084	1102	1.02

M_n: number-average Molecular Weight, M_w: weight-average Molecular Weight, PD = M_w/M_n.

3.3. Influence of Reaction Conditions

To determine the optimal reaction conditions for [HDBU][Im], various factors including temperature, reaction time, catalyst amount, and UV radiation intensity were explored. The effect of the reaction temperature is shown in Figure 2a. Below 150 °C, BHET is barely detectable. As the temperature increases, the conversion rate of PET and yield of BHET increase sharply. From 180 °C to 195 °C, the conversion rate was maintained at 100% and the BHET yield remained steady up to 78.96% from 185 °C to 195 °C. It was obvious that the transesterification reaction is an endothermic reaction, and the reaction temperature played a crucial role in the degradation of PET, so the increase in temperature was conducive to the progress of the reaction. The BHET yields at 185 °C and 195 °C were 78.82% and 78.96%, respectively. At these two temperatures, there is no significant difference in the yield of BHET. It is worth noting that as the temperature rises, so does the energy consumption. Therefore, in terms of energy conservation, 185 °C is the optimal temperature.

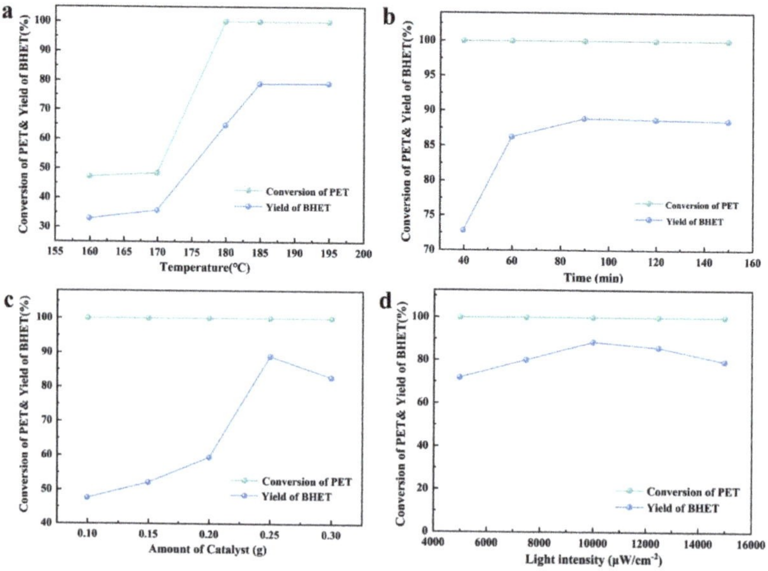

Figure 2. Effect of temperature (**a**) (reaction conditions: 5 g PET, 5 wt% catalyst/PET ratio, 1/4 PET/EG weight ratio with 5000 $\mu W \cdot cm^{-2}$ UV radiation for 40 min), time (**b**) (reaction conditions: 5 g PET, 5 wt% catalyst/PET ratio, 1/4 PET/EG weight ratio with 5000 $\mu W \cdot cm^{-2}$ UV radiation at 185 °C), catalyst dosage (**c**) (reaction conditions: 5 g PET, 1/4 PET/EG weight ratio with 5000 $\mu W \cdot cm^{-2}$ UV radiation at 185 °C for 90 min), and UV radiation intensity (**d**) (reaction conditions: 5 g PET, 5 wt% catalyst/PET ratio, 1/4 PET/EG weight ratio at 185 °C for 90 min) on PET degradation.

The effect of the reaction time is shown in Figure 2b. It is observed that from 40 min to 150 min, the conversion of PET was maintained at 100%. Under UV radiation, the PET conversion reached 100% within 40 min. Degradation was quicker than in conventional experimental conditions [21]. In this process, the conversion of PET reached 100%, while the yield of BHET demonstrated a notable increase as the reaction time was prolonged from 40 min to 90 min. Therefore, we speculated that PET was first degraded into oligomers and then the oligomers turned into BHET. Furthermore, the glycolysis of PET is a reversible reaction, with a dynamic equilibrium existing between the BHET and the oligomers. The reaction reached equilibrium after 90 min; further extending the reaction time did not enhance the yield of BHET. Thus, 90 min was determined as optimal.

Figure 2c illustrates the effect of the amount of catalyst. PET is difficult to degrade without a catalyst. As the [HDBU]Im increased, the PET conversion remained at 100%. However, as the [HDBU]Im dosage increased from 0.1 g to 0.25 g, the yield of BHET increased significantly. The BHET yield attained its maximum value when the catalyst dosage reached 0.25 g, and the BHET yield decreased marginally with further increase of catalyst dosage. This may be due to the excessive amount of catalyst boosting the reverse reaction, causing more BHET to be converted into oligomers. Therefore, excessive catalyst dosage is not conducive to the acquisition of BHET; the optimized catalyst amount is 0.25 g.

Figure 2d illustrates the effect of UV radiation intensity. As the UV radiation intensity increased from 5000 $\mu W \cdot cm^{-2}$ to 15,000 $\mu W \cdot cm^{-2}$, the PET conversion remained at 100%. However, UV radiation intensity can markedly affect BHET yield. When the intensity was raised from 5000 $\mu W \cdot cm^{-2}$ to 10,000 $\mu W \cdot cm^{-2}$, the BHET yield increased gradually. Based on the experimental results above, it can be deduced that UV radiation has the capability to facilitate the breakage of PET molecular chains, thereby accelerating the decomposition of PET into oligomers and monomers. After that, as the intensity was further increased, the yield of BHET decreased. It has been previously suggested that photo-grafting polymerization can occur under UV radiation for compounds containing carbonyl or ester groups [37]. When the intensity of UV radiation exceeds 10,000 μW·cm^{-2}, it has the potential to polymerize BHET into oligomers [38], leading to a decrease in the yield of BHET. Therefore, 10,000 μW·cm^{-2} is the optimal UV radiation intensity, with a BHET yield of 88.9%.

3.4. *Recycling of Solvent and Catalyst*

Recycling of the remaining EG and catalyst is an important issue for environmental and economic reasons. The recycling ability of the catalyst is also crucial for industrialization. It was tested under optimal experimental conditions (5 g PET, 0.25 g IL, 20 g EG, 90 min, 185 °C, 10,000 μW·cm^{-2}). The recycling results are shown in Figure 3, which demonstrates that [HDBU][Im] could sustain its catalytic activity when reused up to six times. After recycling seven times, the catalytic performance decreased, which may be due to the loss of [HDBU][Im] catalyst. As shown in Figure S8, the TGA curve shows that the thermal decomposition temperature of [HDBU][Im] is 201 °C. The reaction temperature of [HDBU][Im]-catalyzed PET degradation is below 201 °C, but prolonged exposure to this temperature may result in partial decomposition of the [HDBU][Im]. Therefore, the catalytic activity of [HDBU][Im] decreased after it was recycled seven times.

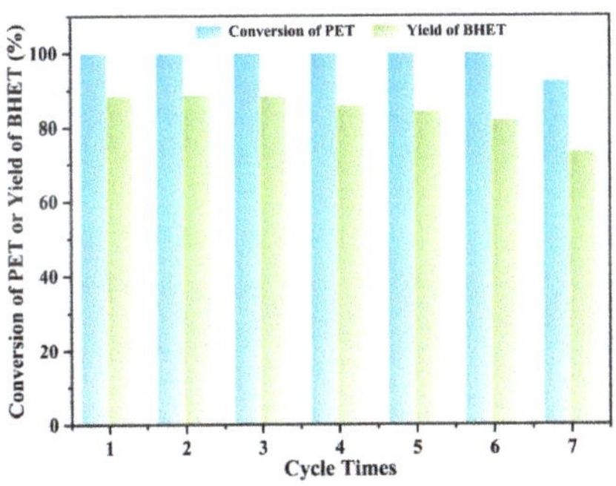

Figure 3. Recycling of [HDBU][Im].

3.5. *Kinetics of PET Glycolysis*

Kinetics can reveal reaction processes [39]. Based on PET conversion, the kinetic of PET is first order [40–42]. We used the shrinking core model to calculate the activation energy of the [HDBU]Im-catalyzed PET glycolysis reaction system. To better understand the effect of UV radiation, kinetics studies with and without UV radiation were all investigated.

Figure 4a demonstrates the effect of temperature on k under UV radiation. A rapid increase in the reaction rate constant is observed as the temperature increases from 170 °C to 185 °C, proving that temperature is an important factor. Calculated using the Arrhenius equation, the relationship between lnk and temperature is shown in Figure 4b. A fine straight line was obtained with correlation coefficient squares larger than 0.98. It indicated that PET glycolysis conforms to a first-order kinetic model. Thus, the activation energy (Ea) is 113.76 kJ·mol^{-1}.

Figure 5a displays the influence of temperature on k without UV radiation, and Figure 5b expresses the relationship between lnk and temperature based on the Arrhenius equation. A fine straight line was obtained with correlation coefficient squares larger than 0.98. The Ea is 166.50 kJ·mol^{-1}. It is evident that UV radiation can reduce the Ea from 166.50 kJ·mol^{-1} to 113.82 kJ·mol^{-1}. The Ea is defined as the minimum amount of energy necessary to initiate a chemical reaction [43]. The lower value of the Ea means the more viable a reaction is and the faster the reaction process. UV radiation can significantly reduce the activation energy of [HDBU][Im]-catalyzed PET glycolysis. Therefore, 100% conversion of PET can be achieved in a shorter time, and the yield of BHET can also be increased.

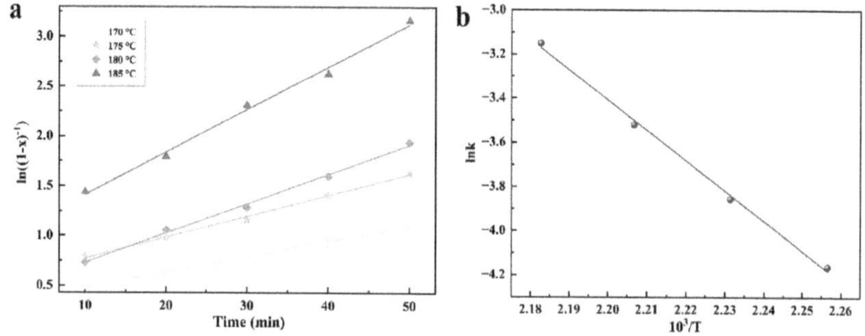

Figure 4. Effect of the temperature on the rate of glycolysis of PET (under UV radiation) (**a**). Arrhenius plot of the rate constant of glycolysis of PET (under UV radiation) (**b**).

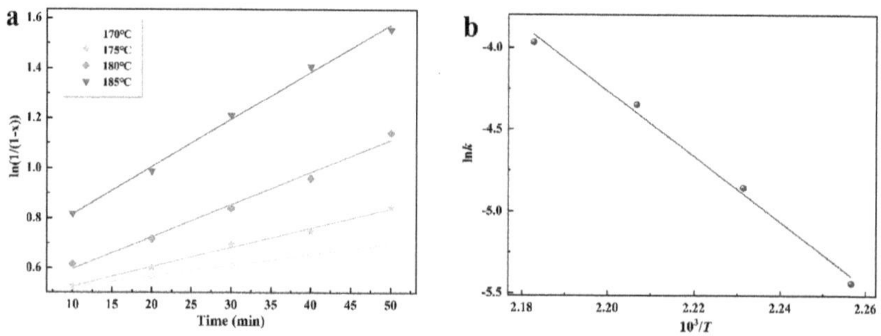

Figure 5. Effect of the temperature on the rate of glycolysis of PET (without UV radiation) (**a**). Arrhenius plot of the rate constant of glycolysis of PET (without UV radiation) (**b**).

3.6. PET Degradation Mechanism

To explore the catalytic mechanism of [HDBU]Im, a series of experiments were conducted under optimal conditions. The results are shown in Table 3. Obviously, the catalytic activity of [HDBU]Im is higher than imidazole or DBU. It can be inferred that $[HDBU]^+$ and $[Im]^-$ play a synergistic role.

Table 3. Catalytic activity comparison among imidazole, DBU, and [HDBU]Im [1].

Catalyst	Conversion of PET (%)	Yield of BHET (%)
Imidazole	22.5	14.5
DBU	100	71.5
[HDBU]Im	100	88.9

[1] Reaction conditions: 5.0 g PET, 0.25 g catalyst, 20.0 g EG reacted for 90 min at 185 °C under 10,000 $\mu W \cdot cm^{-2}$ UV radiation.

According to the literature, some catalysts, such as [Ch][OAc], [dimim]$[FeCl_4]$, [Ch][Gly], and $[Ch]_3[PO_4]$, can form hydrogen bonds (H-bonds) with the hydroxyl hydrogen of EG [24,25,44,45]. The H-bonds elongate the O-H bond of EG hydroxyl and enhance the electronegativity of EG oxygen [46]. Hence, EG can easily attack the carbon of the ester group in PET [21]. To reveal the interaction between [HDBU]Im and EG, FT-IR characterization was conducted for [HDBU]Im/EG mixtures (Figure 6). The O-H vibration of EG showed an obvious 14 cm^{-1} red shift as the [HDBU]Im was introduced into EG. This indicates that [HDBU]Im and EG can form H-bonds, which can effectively activate the hydroxyl group of EG. Therefore, the ability of EG to attack the ester groups in PET is enhanced. To further realize the interaction between EG and catalysts, DBU/EG mixtures were also analyzed (Figure S9). The weight ratios of DBU/EG mixtures are consistent with [HDBU]Im/EG. For

DBU/EG mixtures, the red shift of EG hydroxyl vibration was 8 cm^{-1}. This result further indicates that the hydrogen bonding between [HDBU][Im] and EG is stronger. Therefore, [HDBU]Im could efficiently catalyze the degradation of PET.

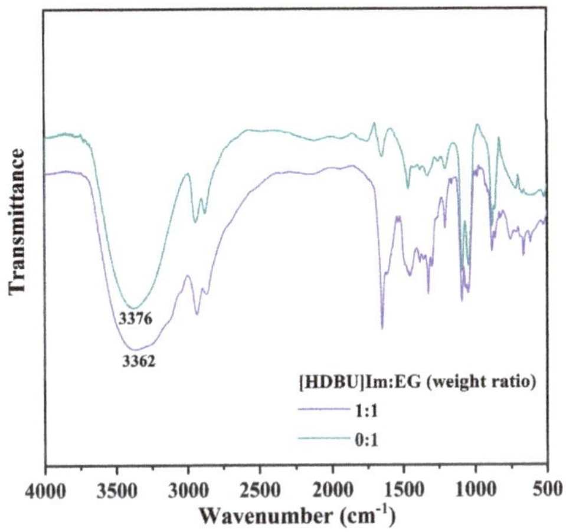

Figure 6. FT-IR spectra of [HDBU]Im/EG mixtures.

To verify the interaction between UV radiation and PET, it is essential to study the UV absorption properties of PET (Figure 7). One strong absorption band appears at nearly 280 nm, corresponding to the benzene ring in PET [47]. The other absorption band appears at nearly 310 nm and corresponds to the carbonyl group in PET [48]. The UV absorption of the carbonyl group causes $n \rightarrow \pi^*$electronic transitions. The transition responsible for UV absorption in the carbonyl group can be traced to the lone pair of electrons on the O atom. One of the electrons in a lone pair can be excited to an empty π^* orbital of the carbonyl group [49], so the electronegativity of the oxygen of C=O in PET becomes higher. Meanwhile, the electrophilicity of the carbon of the C=O group in PET is enhanced. It makes the carbonyl group more vulnerable to EG attacks. As a result, the effect of UV radiation is to activate the carbonyl group, which can promote PET glycolysis. The activation effect is verified by the experimental results in Section 3.5.

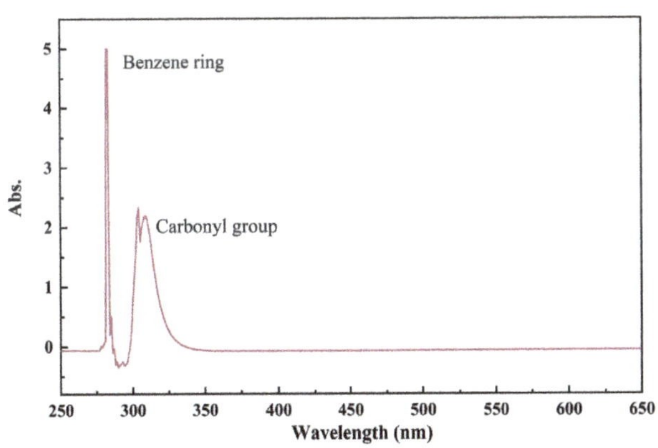

Figure 7. The UV-Visible Spectrum of PET.

In addition, the crystallinity of PET during the reaction process was investigated (Table S1), and it was found that as the degradation reaction progressed, the crystallinity of the residual PET gradually increased. This may be due to the loose arrangement of PET molecular chains in the amorphous region, which is more prone to degradation. The experimental results provide the basis for proposing the potential mechanism of PET glycolysis catalyzed by [HDBU]Im, as illustrated in Scheme 2. EG is activated by [HDBU]Im via H-bonds. The formed H-bonds make the O-H bond of EG hydroxyl longer and the electronegativity of EG oxygen higher. These cause hydrogen to be lost more easily and enhance the nucleophilicity of oxygen. Consequently, it facilitates the attack on the carbon of the PET ester group. DBU has shown relatively high catalytic activity [27]. When DBU is used as catalyst alone, it can activate the carbonyl group of the ester in PET [27]. Hence, $[HDBU]^+$ can activate PET by protonating the carbon of the carbonyl group. Meanwhile, the carbonyl group in PET is also activated by UV radiation. Therefore, PET is more electrophilic, and the highly nucleophilic EG oxygen attacks the carbon of the PET ester group more easily. As the oxygen in the EG hydroxyl attacks the carbon of the ester group in PET, a tetrahedral intermediate will be formed. Thereafter, the hydrogen in EG dissipates. The acyl–oxygen bonds cleave, resulting in the detachment of the $-OCH_2CH_2-$ group, which subsequently combines with H^+ to form $HOCH_2CH_2-$. These processes are repeated to form oligomers and BHET monomers. Finally, chemical equilibrium is formed between the oligomer and the BHET.

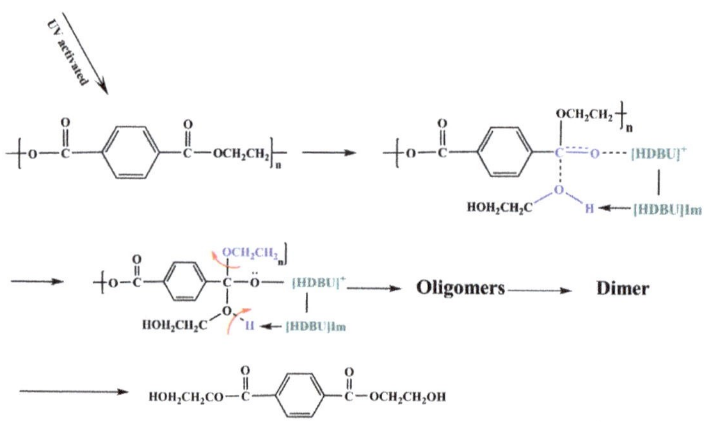

Scheme 2. Mechanism of the glycolysis of PET catalyzed by [HDBU]Im.

4. Conclusions

In summary, we synthesized a series of nonmetallic dibasic ILs, among which [HDBU]Im showed the best performance in catalyzing PET glycolysis. The impact of UV radiation on the PET degradation was further investigated. It was discovered that the utilization of UV radiation expedited the reaction time and enhanced the yield of the monomer BHET. Afterward, the effects of different reaction conditions on PET degradation were investigated; the optimized experimental conditions were 5 g PET, 20 g EG, 0.25 g [HDBU]Im, 185 °C, 10,000 $\mu W \cdot cm^{-2}$ UV radiation reacted for 90 min, and the PET conversion and BHET yield were 100% and 88.9%, respectively. Based on reaction kinetics, UV radiation can reduce the Ea from 166.50 $kJ \cdot mol^{-1}$ to 113.76 $kJ \cdot mol^{-1}$, so the rate of PET glycolysis is considerably enhanced. Throughout the reaction, UV radiation can improve the electronegativity of oxygen in C=O, so the electrophilicity of carbon in carbonyl can be enhanced. This makes PET more vulnerable to EG attacks, accelerating the catalytic reaction rate.

This work verifies the high catalytic performance of dibasic nonmetallic ILs catalyst and UV radiation. In follow-up research, UV radiation can be considered for other reaction systems of PET degradation, aiming to reduce the reaction temperature further, achieve efficient degradation of PET under mild conditions, and provide guidance for the industrialization of PET recycling.

Supplementary Materials: The following supporting information can be downloaded at: https://www.mdpi.com/article/10.3390/ma17071583/s1, Figure S1: FT-IR spectrum of [HDBU][Im] [35,50,51]; Figure S2: ^{1}H NMR spectrum of [HDBU][Im] [34,52]; Figure S3: ^{1}H NMR spectrum of BHET [16]; Figure S4: HPLC spectrum of BHET; Figure S5: TGA curves of PET raw material and BHET [16]; Figure S6: DSC curves of BHET [21]; Figure S7: FTIR spectrum of BHET [22,40,53]; Figure S8: The TGA curve of [HDBU][Im]; Figure S9: FT-IR spectra of DBU/EG mixtures; Table S1: The changes in PET crystallinity during the degradation reaction [54–58].

Author Contributions: Conceptualization, X.L.; Data curation, R.Z. and X.Z.; Formal analysis, R.Z.; Funding acquisition, J.X. (Junli Xu), Q.Z., J.X. (Jiayu Xin), D.Y. and X.L.; Investigation, R.Z. and X.Z.; Methodology, R.Z.; Project administration, X.L.; Resources, X.L.; Supervision, J.X. (Jiayu Xin); Validation, R.Z., X.Z. and X.C.; Visualization, D.Y.; Writing—original draft, R.Z. and X.Z.; Writing—review & editing, X.C., J.X. (Junli Xu), Y.L. and Q.Z. All authors have read and agreed to the published version of the manuscript.

Funding: This research was funded by the Key Program of National Natural Scientific Fund of China (No. 22238011), the National Natural Science Foundation of China (No. 22178343, 22178364, 22178358, 22378406), the Strategic Priority Research Program of Chinese Academy of Science (No. XDA29040700), International Partnership Program of Chinese Academy of Sciences (No. 2023VMA0001), the Youth Innovation Promotion Association of the Chinese Academy of Sciences (No. 2022048), and Beijing Natural Science Foundation (No. 2232030).

Institutional Review Board Statement: Not applicable.

Informed Consent Statement: Not applicable.

Data Availability Statement: Data are contained within the article and Supplementary Materials.

Conflicts of Interest: The authors declare no conflicts of interest.

References

1. Bartolome, L.; Imran, M.; Cho, B.G.; Al-Masry, W.A.; Kim, D.H. Recent developments in the chemical recycling of PET. *Mater. Recycl. Trends Perspect.* **2012**, *406*, 576–596.
2. Zekriardehani, S.; Joshi, A.; Jabarin, S.; Gidley, D.; Coleman, M. Effect of dimethyl terephthalate and dimethyl isophthalate on the free volume and barrier properties of poly (ethylene terephthalate)(PET): Amorphous PET. *Macromolecules* **2018**, *51*, 456–467. [CrossRef]
3. Bartolome, L.; Imran, M.; Lee, K.G.; Sangalang, A.; Ahn, J.K. Superparamagnetic γ-Fe_2O_3 nanoparticles as an easily recoverable catalyst for the chemical recycling of PET. *Green Chem.* **2014**, *16*, 279–286. [CrossRef]
4. Choudhary, K.; Sangwan, K.S.; Goyal, D. Environment and economic impacts assessment of PET waste recycling with conventional and renewable sources of energy. *Procedia CIRP* **2019**, *80*, 422–427. [CrossRef]
5. Palm, G.J.; Reisky, L.; Böttcher, D.; Müller, H.; Michels, E.A.; Walczak, M.C.; Berndt, L.; Weiss, M.S.; Bornscheuer, U.T.; Weber, G. Structure of the plastic-degrading Ideonella sakaiensis MHETase bound to a substrate. *Nat. Commun.* **2019**, *10*, 1717. [CrossRef] [PubMed]
6. Allen, N.S.; Edge, M.; Mohammadian, M.; Jones, K. Physicochemical aspects of the environmental degradation of poly (ethylene terephthalate). *Polym. Degrad. Stab.* **1994**, *43*, 229–237. [CrossRef]
7. Thiounn, T.; Smith, R.C. Advances and approaches for chemical recycling of plastic waste. *J. Polym. Sci.* **2020**, *58*, 1347–1364. [CrossRef]
8. Shojaei, B.; Abtahi, M.; Najafi, M. Chemical recycling of PET: A stepping-stone toward sustainability. *Polym. Adv. Technol.* **2020**, *31*, 2912–2938. [CrossRef]
9. Kathalewar, M.; Dhopatkar, N.; Pacharane, B.; Sabnis, A.; Raut, P.; Bhave, V. Chemical recycling of PET using neopentyl glycol: Reaction kinetics and preparation of polyurethane coatings. *Prog. Org. Coat.* **2013**, *76*, 147–156. [CrossRef]
10. Shirazimoghaddam, S.; Amin, I.; Faria Albanese, J.A.; Shiju, N.R. Chemical recycling of used PET by glycolysis using niobia-based catalysts. *ACS Eng. Au* **2023**, *3*, 37–44. [CrossRef]
11. Arias, J.J.R.; Thielemans, W. Instantaneous hydrolysis of PET bottles: An efficient pathway for the chemical recycling of condensation polymers. *Green Chem.* **2021**, *23*, 9945–9956. [CrossRef]
12. Laldinpuii, Z.; Khiangte, V.; Lalhmangaihzuala, S.; Lalmuanpuia, C.; Pachuau, Z.; Lalhriatpuia, C.; Vanlaldinpuia, K. Methanolysis of PET waste using heterogeneous catalyst of bio-waste origin. *J. Polym. Environ.* **2022**, *30*, 1600–1614. [CrossRef]
13. Mora-Cortes, L.F.; Rivas-Muñoz, A.N.; Neira-Velázquez, M.G.; Contreras-Esquivel, J.C.; Roger, P.; Mora-Cura, Y.N.; Soria-Arguello, G.; Bolaina-Lorenzo, E.D.; Reyna-Martínez, R.; Zugasti-Cruz, A. Biocompatible enhancement of poly (ethylene terephthalate)(PET) waste films by cold plasma aminolysis. *J. Chem. Technol. Biotechnol.* **2022**, *97*, 3001–3010. [CrossRef]

14. Mishra, S.; Goje, A. Kinetics of glycolysis of poly (ethylene terephthalate) waste powder at moderate pressure and temperature. *J. Appl. Polym. Sci.* **2003**, *87*, 1569–1573. [CrossRef]
15. Huang, J.; Veksha, A.; Chan, W.P.; Giannis, A.; Lisak, G. Chemical recycling of plastic waste for sustainable material management: A prospective review on catalysts and processes. *Renew. Sustain. Energy Rev.* **2022**, *154*, 111866. [CrossRef]
16. Imran, M.; Al-Masry, W.A.; Mahmood, A.; Hassan, A.; Haider, S.; Ramay, S.M. Manganese-, cobalt-, and zinc-based mixed-oxide spinels as novel catalysts for the chemical recycling of poly (ethylene terephthalate) via glycolysis. *Polym. Degrad. Stab.* **2013**, *98*, 904–915. [CrossRef]
17. Shukla, S.; Palekar, V.; Pingale, N. Zeolite catalyzed glycolysis of poly (ethylene terephthalate) bottle waste. *J. Appl. Polym. Sci.* **2008**, *110*, 501–506. [CrossRef]
18. Zhu, M.; Li, S.; Li, Z.; Lu, X.; Zhang, S. Investigation of solid catalysts for glycolysis of polyethylene terephthalate. *Chem. Eng. J.* **2012**, *185*, 168–177. [CrossRef]
19. Barnard, E.; Arias, J.J.R.; Thielemans, W. Chemolytic depolymerisation of PET: A review. *Green Chem.* **2021**, *23*, 3765–3789. [CrossRef]
20. Wang, H.; Li, Z.; Liu, Y.; Zhang, X.; Zhang, S. Degradation of poly (ethylene terephthalate) using ionic liquids. *Green Chem.* **2009**, *11*, 1568–1575. [CrossRef]
21. Wang, Q.; Geng, Y.; Lu, X.; Zhang, S. First-row transition metal-containing ionic liquids as highly active catalysts for the glycolysis of poly (ethylene terephthalate)(PET). *ACS Sustain. Chem. Eng.* **2015**, *3*, 340–348. [CrossRef]
22. Wang, Q.; Lu, X.; Zhou, X.; Zhu, M.; He, H.; Zhang, X. 1-Allyl-3-methylimidazolium halometallate ionic liquids as efficient catalysts for the glycolysis of poly (ethylene terephthalate). *J. Appl. Polym. Sci.* **2013**, *129*, 3574–3581. [CrossRef]
23. Yue, Q.F.; Yang, H.G.; Zhang, M.L.; Bai, X.F. Metal-containing ionic liquids: Highly effective catalysts for degradation of poly (ethylene terephthalate). *Adv. Mater. Sci. Eng.* **2014**, *2014*, 454756. [CrossRef]
24. Liu, Y.; Yao, X.; Yao, H.; Zhou, Q.; Xin, J.; Lu, X.; Zhang, S. Degradation of poly (ethylene terephthalate) catalyzed by metal-free choline-based ionic liquids. *Green Chem.* **2020**, *22*, 3122–3131. [CrossRef]
25. Sun, J.; Liu, D.; Young, R.P.; Cruz, A.G.; Isern, N.G.; Schuerg, T.; Cort, J.R.; Simmons, B.A.; Singh, S. Solubilization and Upgrading of High Polyethylene Terephthalate Loadings in a Low-Costing Bifunctional Ionic Liquid. *ChemSusChem* **2018**, *11*, 781–792. [CrossRef] [PubMed]
26. Jehanno, C.; Pérez-Madrigal, M.M.; Demarteau, J.; Sardon, H.; Dove, A.P. Organocatalysis for depolymerisation. *Polym. Chem.* **2019**, *10*, 172–186. [CrossRef]
27. Alnaqbi, M.A.; Mohsin, M.A.; Busheer, R.M.; Haik, Y. Microwave assisted glycolysis of poly (ethylene terephthalate) catalyzed by 1-butyl-3-methylimidazolium bromide ionic liquid. *J. Appl. Polym. Sci.* **2015**, *132*, 41666. [CrossRef]
28. Imran, M.; Kim, B.-K.; Han, M.; Cho, B.G. Sub-and supercritical glycolysis of polyethylene terephthalate (PET) into the monomer bis (2-hydroxyethyl) terephthalate (BHET). *Polym. Degrad. Stab.* **2010**, *95*, 1686–1693. [CrossRef]
29. Le, N.H.; Ngoc Van, T.T.; Shong, B.; Cho, J. Low-temperature glycolysis of polyethylene terephthalate. *ACS Sustain. Chem. Eng.* **2022**, *10*, 17261–17273. [CrossRef]
30. Oliveira, J.; Belchior, A.; da Silva, V.D.; Rotter, A.; Petrovski, Ž.; Almeida, P.L.; Lourenço, N.D. Gaudêncio SP: Marine environmental plastic pollution: Mitigation by microorganism degradation and recycling valorization. *Front. Mar. Sci.* **2020**, *7*, 567126. [CrossRef]
31. Gewert, B.; Plassmann, M.M.; MacLeod, M. Pathways for degradation of plastic polymers floating in the marine environment. *Environ. Sci. Process. Impacts* **2015**, *17*, 1513–1521. [CrossRef] [PubMed]
32. Lee, C.O.; Chae, B.; Kim, S.B.; Jung, Y.M.; Lee, S.W. Two-dimensional correlation analysis study of the photo-degradation of poly (ethylene terephthalate) film. *Vib. Spectrosc.* **2012**, *60*, 142–145. [CrossRef]
33. More, A.P.; Kokate, S.R.; Rane, P.C.; Mhaske, S.T. Studies of different techniques of aminolysis of poly (ethylene terephthalate) with ethylenediamine. *Polym. Bull.* **2017**, *74*, 3269–3282. [CrossRef]
34. Zhu, X.; Song, M.; Xu, Y. DBU-based protic ionic liquids for CO2 capture. *ACS Sustain. Chem. Eng.* **2017**, *5*, 8192–8198. [CrossRef]
35. Zhang, S.; Wang, J.; Liu, X.; Wang, K.; Zhang, C.; Song, H.; Guo, Z.; Lv, Z. Synthesis of 2, 2, 4-trimethyl-1, 3-pentaerediol monoisobutyrate catalyzed by homogeneous catalysis-liquid/liquid separation catalytic system based on Bibasic sites Ionic Liquids. *Appl. Organomet. Chem.* **2020**, *34*, e5915. [CrossRef]
36. Son, S.-Y.; Jo, A.Y.; Jung, G.Y.; Chung, Y.-S.; Lee, S. Accelerating the stabilization of polyacrylonitrile fibers by UV irradiation. *J. Ind. Eng. Chem.* **2019**, *73*, 47–51. [CrossRef]
37. Kordoghli, B.; Khiari, R.; Dhaouadi, H.; Belgacem, M.N.; Mhenni, M.F.; Sakli, F. UV irradiation-assisted grafting of poly (ethylene terephthalate) fabrics. *Colloids Surf. A Physicochem. Eng. Asp.* **2014**, *441*, 606–613. [CrossRef]
38. Fang, L.; Jin, W.; Hong, S.; Nan, H. Surface characterization and in vitro blood compatibility of poly (ethylene terephthalate) immobilized with hirudin. *Plasma Sci. Technol.* **2010**, *12*, 235. [CrossRef]
39. Raheem, A.B.; Noor, Z.Z.; Hassan, A.; Abd Hamid, M.K.; Samsudin, S.A.; Sabeen, A.H. Current developments in chemical recycling of post-consumer polyethylene terephthalate wastes for new materials production: A review. *J. Clean. Prod.* **2019**, *225*, 1052–1064. [CrossRef]
40. Al-Sabagh, A.M.; Yehia, F.Z.; Eissa, A.-M.M.; Moustafa, M.E.; Eshaq, G.; Rabie, A.-R.M.; ElMetwally, A.E. Glycolysis of poly (ethylene terephthalate) catalyzed by the Lewis base ionic liquid [Bmim][OAc]. *Ind. Eng. Chem. Res.* **2014**, *53*, 18443–18451. [CrossRef]

41. López-Fonseca, R.; Duque-Ingunza, I.; de Rivas, B.; Flores-Giraldo, L.; Gutiérrez-Ortiz, J.I. Kinetics of catalytic glycolysis of PET wastes with sodium carbonate. *Chem. Eng. J.* **2011**, *168*, 312–320. [CrossRef]
42. Pingale, N.; Shukla, S. Microwave assisted ecofriendly recycling of poly (ethylene terephthalate) bottle waste. *Eur. Polym. J.* **2008**, *44*, 4151–4156. [CrossRef]
43. Lu, X.; Xu, R.; Sun, K.; Jiang, J.; Sun, Y.; Zhang, Y. Study on the effect of torrefaction on pyrolysis kinetics and thermal behavior of cornstalk based on a combined approach of chemical and structural analyses. *ACS Omega* **2022**, *7*, 13789–13800. [CrossRef] [PubMed]
44. Scé, F.; Cano, I.; Martin, C.; Beobide, G.; Castillo, O.; de Pedro, I. Comparing conventional and microwave-assisted heating in PET degradation mediated by imidazolium-based halometallate complexes. *New J. Chem.* **2019**, *43*, 3476–3485. [CrossRef]
45. Marullo, S.; Rizzo, C.; Dintcheva, N.T.; D'Anna, F. Amino acid-based cholinium ionic liquids as sustainable catalysts for PET depolymerization. *ACS Sustain. Chem. Eng.* **2021**, *9*, 15157–15165. [CrossRef]
46. Ju, Z.; Xiao, W.; Lu, X.; Liu, X.; Yao, X.; Zhang, X.; Zhang, S. Theoretical studies on glycolysis of poly (ethylene terephthalate) in ionic liquids. *RSC Adv.* **2018**, *8*, 8209–8219. [CrossRef] [PubMed]
47. Takeuchi, M.; Hidaka, M.; Anpo, M. Simple evaluation of the adsorption states of benzene molecule on the hydroxyl, H+ and Na+ sites of Y-zeolite surfaces by using UV absorption spectroscopy. *Res. Chem. Intermed.* **2014**, *40*, 2315–2325. [CrossRef]
48. Liu, F.; Beames, J.M.; Green, A.M.; Lester, M.I. UV spectroscopic characterization of dimethyl-and ethyl-substituted carbonyl oxides. *J. Phys. Chem. A* **2014**, *118*, 2298–2306. [CrossRef] [PubMed]
49. Atkins, P.W.; De Paula, J.; Keeler, J. *Atkins' Physical Chemistry*; Oxford University Press: Oxford, UK, 2023.
50. Lv, Z.; Zhang, S.; Guo, Z.; Cheng, X.; Wang, J.; Zhang, C. Synthesis of alcohol ester 12 in 1, 8-diazabicyclo [5.4.0] undec-7-ene (DBU)-based Self-separation catalytic system. *Appl. Organomet. Chem.* **2019**, *33*, e5145. [CrossRef]
51. Simons W, W. *Sadtler Handbook of Infrared Spectra*; Sadtler Research Laboratories: Watford, UK, 1978.
52. Liu, M.; Guo, J.; Gu, Y.; Gao, J.; Liu, F. Versatile Imidazole-Anion-Derived Ionic Liquids with Unparalleled Activity for Alcoholysis of Polyester Wastes under Mild and Green Conditions. *ACS Sustain. Chem. Eng.* **2018**, *6*, 15127–15134. [CrossRef]
53. Fang, P.; Liu, B.; Xu, J.; Zhou, Q.; Zhang, S.; Ma, J. High-efficiency glycolysis of poly(ethylene terephthalate) by sandwich-structure polyoxometalate catalyst with two active sites. *Polym. Degrad. Stab.* **2018**, *156*, 22–31. [CrossRef]
54. Matsuo, M.; Luo, Y.; Galeski, A. Gauche-trans transitions in amorphous polymers under annealing: Lattice model and polarized light scattering. *Phys. Rev. E* **2009**, *79*, 041801. [CrossRef] [PubMed]
55. Zhang, Y.; Lu, Y.; Duan, Y.; Zhang, J.; Yan, S.; Shen, D. Reflection–absorption infrared spectroscopy investigation of the crystallization kinetics of poly (ethylene terephthalate) ultrathin films. *J. Polym. Sci. Part B* **2004**, *42*, 4440–4447. [CrossRef]
56. Cole, K.C.; Ajji, A.; Pellerin, E. New insights into the development of ordered structure in poly (ethylene terephthalate). 1. Results from external reflection infrared spectroscopy. *Macromolecules* **2002**, *35*, 770–784. [CrossRef]
57. Wang, Y.; Shen, D.; Qian, R. Subglass-transition-temperature annealing of poly (ethylene terephthalate) studied by FTIR. *J. Polym. Sci. Part B* **1998**, *36*, 783–788. [CrossRef]
58. Qian, R.; Shen, D.; Sun, F.; Wu, L. The effects of physical ageing on conformational changes of poly (ethylene terephthalate) in the glass transition region. *Macromol. Chem. Phys.* **1996**, *197*, 1485–1493. [CrossRef]

materials

MDPI

Article

Development of a Recycling Process and Characterization of EVA, PVDF, and PET Polymers from End-of-Life PV Modules

Marek Królikowski [1,*], Michał Fotek [2], Piotr Żach [2] and Marcin Michałowski [3]

[1] Faculty of Chemistry, Warsaw University of Technology, Noakowskiego 3, 00-664 Warsaw, Poland
[2] Faculty of Automotive and Construction Machinery, Warsaw University of Technology, Narbutta 84, 02-524 Warsaw, Poland; piotr.zach@pw.edu.pl (P.Ż.)
[3] Faculty of Mechatronics, Warsaw University of Technology, św. Andrzeja Boboli 8, 02-525 Warsaw, Poland; marcin.michalowski@pw.edu.pl
[*] Correspondence: marek.krolikowski@pw.edu.pl; Tel.: +48-22-234-7475

Abstract: Photovoltaic (PV) modules are highly efficient power generators associated with solar energy. The rapid growth of the PV industry will lead to a sharp increase in the waste generated from PV panels. However, electro-waste can be successfully used as a source of secondary materials. In this study, a unique procedure for recycling PV modules was developed. In the first stage, the aluminum frame and junction box, 18wt%. and 1wt%. of the module, respectively, were removed. The following stage was crucial, involving a mechanical–thermal method to remove the glass, which accounts for 70wt%. As a result, only 11wt%. of the initial mass of the PV was subjected to the next stage of chemical delamination, which reduced the amount of solvent used. Toluene was used to swell the ethylene vinyl acetate, EVA, and allow for the separation of the PV module. The effects of temperature and ultrasound on separation time were investigated. After the separation of silicon cells, metal ribbons, EVA, and the backsheet were obtained. The purity of the polymers was determined by FTIR and elemental analysis. Thermal properties were measured using DSC calorimetry to determine the basic parameters of the material.

Keywords: photovoltaic module; recycling PV module; chemical treatment; mechanical–thermal treatment; FTIR spectroscopy; elemental analysis; EVA; PET; PVDF

1. Introduction

The widespread use of traditional fossil fuels, as well as the resultant increased environmental pollution, has prompted a gradual transition in energy supply to renewable energy. Additionally, developing sustainable renewable energy is an effective strategy for reducing carbon emissions. Due to its non-toxic emissions and simple installation, photovoltaic power generation is gradually gaining dominance in renewable energy. The total global cumulative PV installed capacity was 1185 GW at the end of 2022 [1]. Compared to 2021, this is an increase of 240 GW of new systems installed and commissioned. The Chinese market continued to dominate new and cumulative capacity and added 106 GW. Europe demonstrated continued strong growth with 39 GW installed, led by Spain, with 8.1 GW, Germany, with 7.5 GW, Poland, with 4.9 GW, and the Netherlands, with 3.9 GW. The rapid growth of the photovoltaic industry will lead to a sharp increase in the waste that is generated from PV panels. The most common silicon solar cells have a 20–30 year lifespan on average. The amount of PV e-waste will rise to 60–78 million tons globally. Degraded PV modules are predicted to produce 10% of all electronic waste by the year 2050 [2]. However, electro-waste can be successfully used as a source of secondary materials. Recently, several research publications, particularly review papers on the recycling of PV modules, have been published [3–6]. In general, PV recycling involves three stages: (1) manual/mechanical disassembly; (2) delamination: thermal, physical/mechanical, chemical, or combined; and (3) separation into silicon, metal and polymer fractions [7]. To separate specific layers of PV modules, physical procedures, chemical swelling or dissolving processes, and pyrolysis can be applied [8]. The physical and mechanical approaches focus on crushing and sifting the materials of PV modules. The method is straightforward, but it does not achieve sufficient separation of distinct layers, lowering the resource recovery ratio. Because physical and mechanical approaches do not affect EVA characteristics, EVA bonding remains effective. As a result, eliminating EVA bonding to adequately separate the different layers is a precondition for adequately recovering the available resources in PV modules. According to relevant studies [9,10], pyrolysis is the most effective approach for directly eliminating EVA and separating different layers. However, fluorine compounds from the polyvinyl fluoride, PVF, or polyvinylidene fluoride, PVDF, of the backsheet can be released into the environment in a direct thermal process [11,12]. Furthermore, glass and the backsheet inhibit the release of low-molecular-weight organic compounds during EVA pyrolysis, which can result in the contamination of glass and silicon cells with decomposition hydrocarbons and further mixing of glass and solar cells, lowering resource recovery rates. The chemical swelling or dissolving methods appear to be a promising way of separating the different layers of a PV module due to their low energy consumption and high rate of separation [13–15]. In the first step, the tempered glass was recovered using an organic solvent. Additionally, EVA copolymer contaminants were removed from the PV cells by thermal decomposition. On the other hand, silicon was obtained through a chemical etching process using HF and HNO_3 with a surfactant by removing metal impurities from the surface of the recovered PV cell. It was possible to generate a high silicon yield [16]. The backsheet can be removed mechanically or using ethanol or toluene vapor [12,15]. However, the chemical process has disadvantages. EVA typically swells excessively, causing solar cells to be crushed [17]. Organic solvents such as benzene, toluene, o-dichlorobenzene, trichloroethylene, tetrahydrofuran, methyl ethyl ketone, methyl isobutyl ketone, or petroleum benzine are used in the swelling process, but some of them are toxic or mutagenic and contain halogens. The processes require a large amount of solvent and a high temperature, $T > 70$–80 °C, close to the boiling point of the solvent at atmospheric pressure, and last a long time, over 1 h [16–18]. As a consequence, recycling PV modules can be costly and time-consuming.

This study presents an alternative methodology for the separation of PV modules after their end of life. At first, the aluminum frame, junction box, and connecting wires were mechanically separated. The initial phase, while not novel, stands out primarily due to simplicity, affordability, and expediency. Subsequently, within a thermomechanical procedure, a layer of broken tempered glass was separated, which is a crucial stage since glass limits contact of the organic solvent with the next copolymer EVA layer. Moreover, it is worth noting that glass constitutes up to 70wt%. of the PV module, as well as a significant volume. It is pointless to use excess solvent, which does not affect the glass. The next procedure involves a chemical method with toluene as a solvent applied to the delamination of PV cells and the backsheet. The process was carried out at $T = 35$ °C and in a time below $t = 40$ min to obtain a swollen EVA copolymer separated from the silicon cells, busbars, and backsheet. A significantly shorter delamination time for PV modules was obtained in comparison to the procedures described in the literature [19–21]. The organic solvent also caused the backsheet to separate into two distinct layers composed of PVDF and PET. In a further step, the obtained polymers were confirmed and characterized by FTIR spectroscopy, elemental analysis, EA, and differential scanning calorimetry, DSC. This work presents the development of a straightforward approach for the environmentally sustainable recovery of EVA, PET, and PVDF from wasted crystalline silicon PV modules. The recovered materials can be effectively reused in PV modules, as well as find applications in the packaging and textile sectors.

2. Materials and Methods

2.1. Photovoltaic Modules

In order to analyze the various layers of PV modules, PV sheet sections were prepared from standard 275 W commercial end-of-life PV modules (polycrystalline–silicon type) obtained from the commercial PV facility of DAH Solar Co., Ltd., Hefei, China. The first step in recycling PV modules is to remove the aluminum frame and junction box, 18%wt. and 1%wt., respectively. This is usually conducted mechanically, and similar methods are presented in the literature [22–24]. To obtain the lamination structure required for the study, metal pliers and screwdrivers were used in advance to remove the aluminum frame and junction box of the PV module.

In the next step, the glass layer was removed. It should be noted that in most recycled PV modules, the glass sheet is broken into pieces a few to several millimeters in diameter and the glass is not recovered as a whole sheet due to physical damage during the lifetime, the method of dismantling and transporting the PV modules, and damage during the removal of the frame and junction box. The PV module on the glass side was thermally treated using Steinel® HL1920 hot air guns (Steinel, Bloomington, MN, USA) with 2000 W of power, an infinitely variable temperature adjustment from 80 to 600 °C, and airflow control. The PV sample was heated to 170 ± 5 °C, at which point the EVA became soft and glass could be recovered by applying mechanical force. The temperature of the PV module was monitored using a pyrometer Benetech GM700 (Shenzhen Jumaoyuan Science And Technology Co., Ltd., Shenzhen, China) with an accuracy ±1.5 °C. After removing the glass, the laminated sample was cut into 5 cm × 5 cm pieces using scissors. Figure 1 shows photos of the PV module, the resulting PV sheet sections obtained from it that were used during the experiment, and the removed glass.

(a) (b) (c)

Figure 1. A photo of (**a**) an end-of-life PV module, (**b**) laminated PV samples after glass removal, and (**c**) removed glass.

2.2. Swelling of EVA and Separation Process

Prepared in the previous stage, laminated samples were next subjected to solvent treatment to swell and separate EVA, extract the PV cells and connecting wires (busbars and fingers), and separate the backsheet. According to our experimental findings and the available literature [13,16,17], toluene (CAS No. 108-88-3, supplier: Chempur, purity > 99%, Piekary Śląskie, Poland) was used to swell the polymer. The swelling processes were prepared in a jacketed and thermostatted glass vessel with a volume of 150 mL. The jackets were connected to the thermostatic water bath, Julabo CORIO CD-BC6 (JULABO GmbH, Seelbach, Germany), to maintain a constant temperature with an accuracy of $T = 0.05$ °C. The heterogeneous mixtures of laminated samples and solvent were vigorously stirred using a mechanical stirrer IKA Microstar 15 Control (IKA, Staufen, Germany) with an R 1355 centrifugal stirrer. The rotation speed of the stirrer was constant and equal to 500 rpm, which was enough to provide perfect contact between the laminated samples and the solvent. In the next part of the experiment, an ultrasonic bath, PROCLEAN 2.0M ECO (Ulsonicx, Berlin, Germany), with a capacity of 2 dm^3 and ultrasound power of 60 W was used. The bath was filled halfway with distilled water. The delamination process was carried out in a 250 mL glass beaker placed in the bath. During the measurements, the temperature of the solution in the beaker was measured using a P 750 thermometer (Dostmann electronic GmbH, Wertheim, Germany) equipped with a PT100 sensor.

To determine the degree of delamination of PV modules during the EVA copolymer swelling process, 2 cm^3 of solvent was sampled and subjected to analysis of physicochemical properties, including density and dynamic viscosity. The swelling process of the EVA copolymer entails partial dissolution, especially of its shorter, uncrosslinked chains. In addition, some small-molecule compounds derived from the EVA copolymer migrate into the solvent. Consequently, the density and dynamic viscosity of the solvent change during the process of PV module delamination. The dependence of the physicochemical properties of toluene as a function of the degree of PV module delamination is presented in Figures S1 and S2 and Table S1 in the Supplementary Materials.

2.2.1. Density

The density of the toluene solution was determined under ambient pressure using a vibrating tube densimeter—DMA 4500 M, Anton Paar, Graz, Austria. The densimeter is equipped with an automated adjustment mechanism to account for the viscosity of the liquid sample. Calibration was conducted using water that had undergone double distillation and degassing, as well as air that had been dried. The utilization of two integrated Pt 100 platinum thermometers enables the achievement of precise temperature control with an accuracy of 0.05 °C. The measurement uncertainty was estimated to be better than 5×10^{-4} $g{\cdot}cm^{-3}$.

2.2.2. Dynamic Viscosity

The dynamic viscosity was measured using a cone/plate rheometer, model DVNext-LV, AMETEK Brookfield (Middleborough, MA, USA), with a relative uncertainty of $u_r(\eta) = 0.03$ mPa·s. The viscosity error was determined using the standards APN26E and APN75 from Paragon Scientific Ltd. (Prenton, UK), as well as n-tetradecane (Alfa Aesar, Graz, Austria, purity > 99%). All measurements were conducted at an atmospheric pressure of p = 100 kPa, with an uncertainty in the temperature measurement of 0.1 K. The viscosity measurement was conducted at a constant temperature, T = 25.0 °C.

2.3. *Analysis Methods: FT-IR Spectrometry and Elemental Analysis*

The Fourier transform infrared spectrum analyses of the polymers were carried out using a Nicolet iS5, Thermo Scientific (Waltham, MA, USA) Mid Infrared FT-IR spectrometer equipped with an iD7 ATR Optical Base. Immediately before the measurement, the ATR crystal with ethanol was cleaned and dried. Then, the samples of polymers were directly placed on the surface of the ATR. The wavenumber ranged from 3900 to 400 cm^{-1}.

Elemental analysis (EA) was performed on a CHNS Analyzer VARIO EL III (Elementar Analysensysteme GmbH, Langenselbold, Germany).

2.4. *Differential Scanning Calorimetry*

The differential scanning calorimetry (DSC) technique was used to determine the temperature (T_g) and heat capacity change in the glass transition ($\Delta C_{p(g)}$). The studies were carried out using a Mettler Toledo DSC 1 STARe System calorimeter fitted (Mettler Toledo, Toronto, ON, Canada) with a liquid nitrogen cooling system and set to heat-flux mode. The sample cell was constantly fluxed with high-purity nitrogen at a constant flow rate of 20 $mL{\cdot}min^{-1}$. The apparatus was calibrated with the 99.9999 mol% purity indium sample and with high-purity ethylbenzene, n-octane, n-decane, n-octadecane, n-eicosane, cyclohexane, biphenyl, and water. The calibration experiment was carried out with a 5 $°C{\cdot}min^{-1}$ heating rate in the temperature range from −95 °C to 200 °C.

3. Results and Discussion

The external components of the crystalline PV modules consist of the aluminum frame and junction box, which account for approximately 18%wt. of the total PV mass, and the glass, which comprises around 70%wt. The process of removing the frame and junction box can be considered rather straightforward from a mechanical standpoint. Furthermore, a considerable fraction of PV modules that are stored lack these components, primarily because the aluminum and copper materials present in the frame, cables, and junction box hold significant value and are removed before the PV module becomes a waste of electrical and electronic equipment, WEEE. The subsequent stage in the separation process of the photovoltaic (PV) modules, which involves the elimination of the tempered glass layer, holds significant importance. On the one hand, the heaviest fraction, the glass of the PV module, is removed, and on the other hand, the direct exposure and chemical treatment of the EVA lamination layer are enabled.

3.1. Thermal Separation of the Glass Layer

The process of delamination of PV modules at low temperatures, namely, those below the thermal breakdown point of EVA, is a complex and challenging phenomenon. The strong adhesion of the EVA copolymer to glass surfaces is due to the siloxane chemical bonds formed [25]. Silane-based coupling agents are commonly used across several industries as adhesion promoters, facilitating the bonding of organic polymers to inorganic (glass and Si) and organic (PVDF) substrates. These agents have gained recognition as a regular ingredient in encapsulating materials specifically designed for PV module applications [26]. The methoxysilyl groups present in silane undergo hydrolysis, resulting in the formation of silanols. These silanols then undergo condensation on the hydroxylated glass surface, leading to the formation of siloxane bonds. The silane compound also exhibits affinity toward EVA and undergoes polymerization with the polymer itself, resulting in the formation of an interphase region, as depicted in Figure 2.

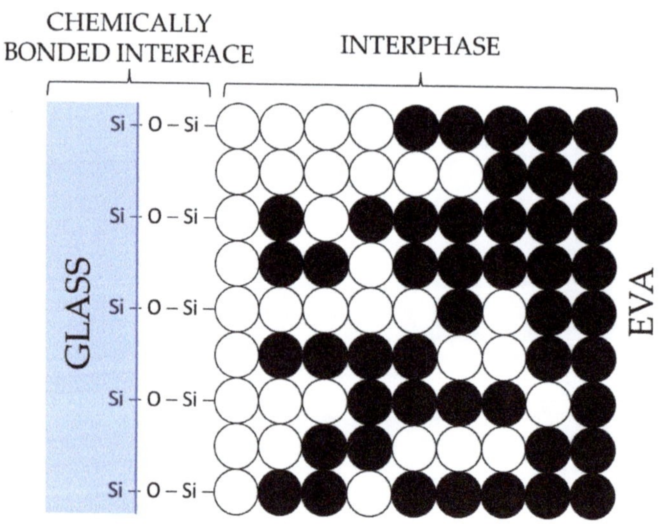

Figure 2. A schematic illustrating a potential interfacial configuration between EVA and glass, considering the inclusion of a silane adhesion promoter [27].

The samples of PV modules on the glass side were thermally treated using hot air guns. The surface of the glass exhibited a temperature range from 160 °C to 170 °C. The glass layer is quite thick, 4 mm, and all of the heat does not penetrate the glass to the polymer surface. A glass temperature of 170 °C ensures that EVA is above the melting point, T_m = 66.9 °C, and below the decomposition temperature, T_d = 215–385 °C, in the air environment [28,29]. Under these conditions, the EVA is soft, and the glass pulls easily from the surface of the PV module. The process of separating the glass required the bending of the PV module and the use of mechanical force, achieved through the utilization of a metal blade to pry apart the glass fragments. Throughout the process, no instances of EVA yellowing or residual gas release were recorded, indicating the absence of polymer degradation. The PV module with separated glass is shown in Figure 1b.

In the process of heating the PV module, it is feasible to remove the outermost layer of the backsheet as well, which is PET. It is possible after heating the backsheet to a temperature above 130 °C and mechanically pulling off the PET layer. Moreover, the separation of the PET layer from the backsheet also occurs when it is exposed to a solvent, as detailed in the next paragraph.

3.2. Chemical Treatment of EVA

A variety of organic solvents were used to swell the EVA copolymer and fractionate the photovoltaic module. The EVA copolymer exhibits the highest degree of swelling and partial dissolution in toluene, cyclohexane, xylene, tetrahydrofuran, o-dichlorobenzene, and trichloroethylene. In this study, toluene was selected due to its popularity, affordability, lower volatility, and reduced harm compared to chlorohydrocarbons. Nevertheless, a combination of xylene isomers exhibits comparable efficacy.

First, the effect of temperature on the rate of the delamination process was investigated. Each time, a sample of a 5 cm × 5 cm PV module and the same amount of 100 mL of solvent were used for the measurement. The system was stirred with a mechanical stirrer at a rotational speed of 500 rpm. Measurements were performed at three distinct temperatures, denoted as T_1 = 25 °C, T_2 = 35 °C, and T_3 = 45 °C. In Figure 3, the dependence of the degree of PV module delamination on the duration of the EVA swelling process is presented.

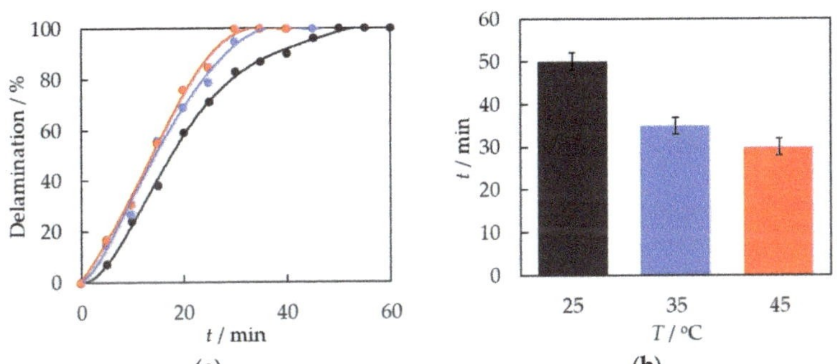

Figure 3. (**a**) The degree of PV module delamination as a function of time, *t*, during solvent exposure. dots—experimental data. A solid line is a guide for the eye. •—T_1 = 25 °C; •—T_2 = 35 °C; •—T_3 = 45 °C. (**b**) Delamination time of PV modules as a function of temperature.

It can be seen from the above data that the effect of temperature on the delamination process of PV modules is significant. For T_1 = 25 °C, the delamination and swelling process of EVA ended after 50 min; for T_2 = 35 °C, after 35 min; and for T_3 = 45 °C, after 30 min. As the temperature increases, the delamination process becomes shorter. At higher temperatures, the process of solvent penetration into the polymer matrix is facilitated. The degree of polymer segment mobility in the solvent phase experiences a significant rise. EVA has a higher degree of swelling efficiency.

In the next step, an ultrasonic bath was used to test the effect of ultrasonics on the swelling process of the EVA copolymer and the delamination of PV modules. A sample of a 5 cm × 5 cm PV module and the same amount of solvent as in previous studies, 100 mL of toluene, were placed in the beaker. The process was carried out in two ways: without mechanical stirring and with 500 rpm mechanical stirring. The process was carried out until constant values of density and viscosity of the solvent solution and delamination of the photovoltaic module were established. Figure 4 shows the results compared to the process carried out at a constant temperature, $T = 35$ °C, with mechanical stirring of 500 rpm.

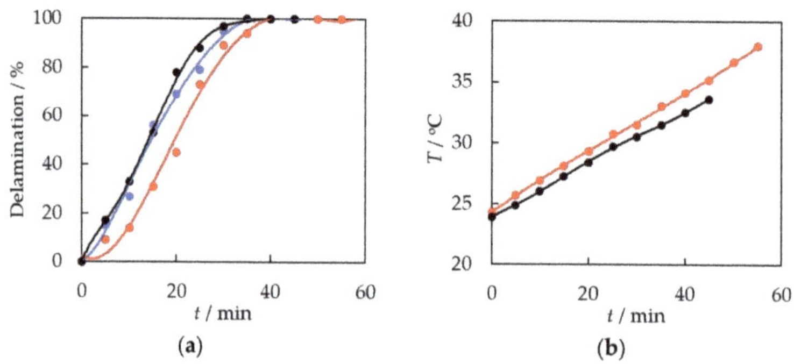

Figure 4. (**a**) The degree of PV module delamination as a function of time, t,, during solvent exposure. dots—experimental data. A solid line is a guide for the eye. •—constant temperature, $T_1 = 25$ °C; stirring at 500 rpm; •—ultrasonic bath without stirring; •—T_3 = ultrasonic bath, stirring at 500 rpm. (**b**) Temperature changes in the ultrasonic bath during the delamination process: •—ultrasonic bath without stirring; •—T_3 = ultrasonic bath, stirring at 500 rpm.

The EVA swelling and delamination of PV modules using the ultrasonic bath without mechanical stirring exhibited a less dynamic and slower behavior compared to the procedure performed at a constant temperature, $T = 35$ °C, with mixing. Behavior is attributed to two factors: first, the lack of stirring makes it more difficult to access the EVA and for the solvent to penetrate it, detaching fragments of the swelling polymer from the surface of the module. Moreover, the delamination process began at a slightly lower temperature of 24.3 °C. As depicted in Figure 4, the temperature exhibited an upward trend throughout the procedure. As the temperature increases, the delamination and swelling process becomes more active. The temperature of the solution in an ultrasonic bath increases in a nearly linear manner.

The most dynamic effect of EVA swelling and PV module delamination was achieved using both mechanical stirring and ultrasound, as shown in Figure 4. Ultrasound is thought to have two effects: firstly, it raises the temperature of the system, making it easier for the EVA to expand. Secondly, it helps the solvent penetrate the polymer chains and further breaks down the swollen EVA copolymer into smaller pieces. A delamination time of 35 min was achieved. The current time is similar to the delamination time achieved in a procedure conducted at a consistent temperature of $T = 35$ °C with agitation at 500 rpm.

The chemical delamination process can be divided into the following distinct steps, which are presented in Figure 5. During the initial stage (stage 1), there is an indication of swelling and fragmentation of the top laminating layer of silicon cells, as shown in Figure 5a. Stage 2 of the procedure entails removing the laminate layer, which causes the metal dust and crushed silicon cells to separate and fall into the solvent solution, as shown in Figure 5b. Suspension of metallic dust particles is obtained. Stage 3 involves the swelling and defragmentation of the inner laminating EVA layer that is bonded to the PVDF polymer. In stage 4, the backsheet undergoes a process of separation into two distinct components, namely, PVDF and PET, presented in Figure 5c,d, respectively. The backsheet separation is facilitated by the presence of a solvent and the application of mechanical forces generated through mixing.

(a) (b) (c) (d)

Figure 5. (**a**) Swollen and fragmented laminating EVA layer, (**b**) metallic dust suspension in the solvent, (**c**) PVDF, and (**d**) PET.

During the chemical delamination process, the following material fractions are obtained: polymers, silicon cells, and metals. From the polymer fraction, the backsheet, separated into PET and PVDF, can be easily isolated from crushed EVA and PV cells with a diameter < 3 mm using a sieve with a mesh diameter of 6–10 mm. Both PET and PVDF exhibit resistance to swelling and dissolution upon exposure to toluene; however, they are separated from each other. PET is obtained in pure form after evaporation of the residual solvent, as will be confirmed by the analyses described in the next section. The purity of separated PVDF is comparatively lower than that of PET. The PVDF surface exhibits a coating of copolymer EVA residues. It occurs as a result of the interpenetration of PVDF and EVA polymer chains during the lamination process, which is conducted at an elevated temperature. The solvent used is unable to dissolve the EVA and separate it from the PVDF surface.

In the next step, the EVA and the crushed PV cells are separated, as illustrated in Figure 5a. Following the evaporation of the residual solvent, the separation of the PV cells from the EVA material occurs due to the disparity in density between these substances. For this purpose, the mixture of EVA and PV cells was placed in a glass beaker with water, as shown in Figure 6b. The density of EVA is lower than that of water and the polymer floats on the surface. A qualitative analysis was performed on the EVA that had been isolated and dried. Some of the recovered and dried PV cells contain a crosslinked EVA laminating polymer on their surface which has not been swollen and separated by etching with the solvent, toluene. The only way to remove the residue of laminate contamination is by pyrolysis or burning [28,29]. Figure 6c shows the PV cells after EVA burns in the flame of a gas burner.

(a) (b) (c)

Figure 6. (**a**) A mixture of EVA and PV cells after chemical treatment, (**b**) the separation of EVA (on the surface) and PV cells (at the bottom of the beaker) in water, and (**c**) PV cells after burning.

3.3. Characteristics of Separated Materials

The recovered polymers were subjected to a range of characterization procedures to assess their structural composition, elemental composition, and thermal behavior.

3.3.1. FTIR Spectroscopy

The chemical compositions of the recovered polymer samples were analyzed using Fourier transform infrared (FTIR) spectroscopy in attenuated total reflectance (ATR) mode. The FTIR results show that the encapsulating material is EVA (Figure 7) and that the backsheet layer is composed of PET (Figure 8) and PVDF (Figure 9). To determine the functional groups present in the recovered EVA, PET, and PVDF polymers, the material was analyzed in the spectral range from 4000 to 400 cm^{-1} and compared to the reference spectra of EVA, PET, and PVDF available in the literature [22,30]. The major vibrations of the EVA and literature sample are described and shown in Figure 6. Two bands at 2888 cm^{-1} and 2848 cm^{-1} are assigned to the symmetric and antisymmetric stretching bond of C–H in the $-CH_2$ group. The band at 1735 cm^{-1} comes from the stretching bond of the carbonyl group, C=O. Two bands are present at 1468 cm^{-1} and 1370 cm^{-1}, which is due to the bending of C−H. The EVA sample has a very concentrated band at 1233 cm^{-1}, which indicates the presence of a C−O−C bond in the ester functional group. Lastly, there are two minor bands observed at 1018 cm^{-1} and 719 cm^{-1}, which are attributed to the stretching of the C−O bond and the rocking vibration of the $-CH_2$ group, respectively.

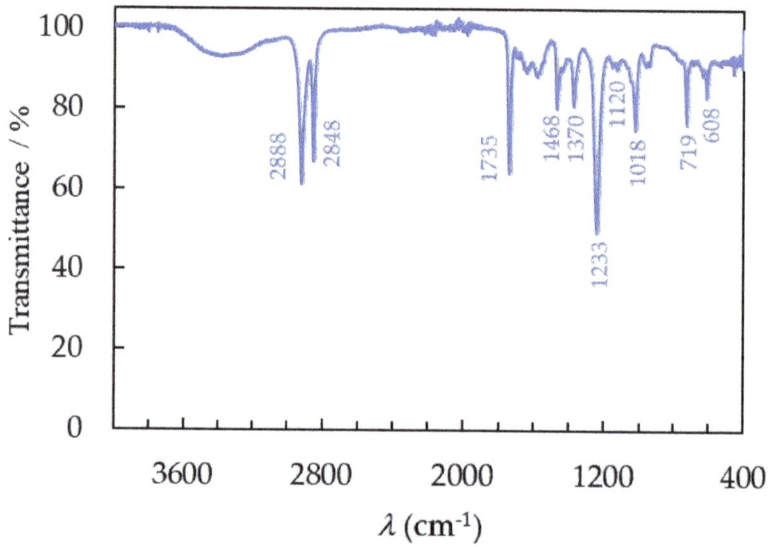

Figure 7. The FTIR spectra of recovered EVA after separation of PV module.

Furthermore, the broad peak between 3200 cm^{-1} and 3600 cm^{-1} corresponds to the vibration of the hydroxyl group −OH. The presence of moisture in the EVA is due to the production of acetic acid during deacetylation [31], as well as moisture permeating the laminating layer during use and the separation process. The FTIR spectrum of the investigated recovered material fully corresponds to the spectrum from the literature data for both the recycled and virgin polymers [30].

In the PET spectra, small bands were observed in the region of 2918 cm^{-1}; these are symmetric and antisymmetric stretching bonds of C−H in the $-CH_2$ group that were also present in the EVA. The larger band observed at 1712 cm^{-1} corresponds to the bond stretching of the carbonyl group, C=O, and at 1340 cm^{-1} to the bending vibrations of C-H bonds in a methylene group. The bands detected at 723 cm^{-1}, 872 cm^{-1}, and 1409 cm^{-1} were ascribed to modifications arising from the CH=CH aromatic ring. Additionally, 1094 cm^{-1} bands associated with symmetric C−O stretching vibrations and 1245 cm^{-1} bands associated with asymmetric C−O−C stretching vibrations in the ester group were identified. The FTIR spectrum of the PET recovered from the backsheet is in complete agreement with the spectrum found in the published data [22]. FTIR analysis confirms that the EVA from the laminating layer and PET recovered from the backsheet, during both the lamination process and the use of PV panels, as well as during the mechanical and chemical recycling processes, have not undergone degradation and have maintained their original structure.

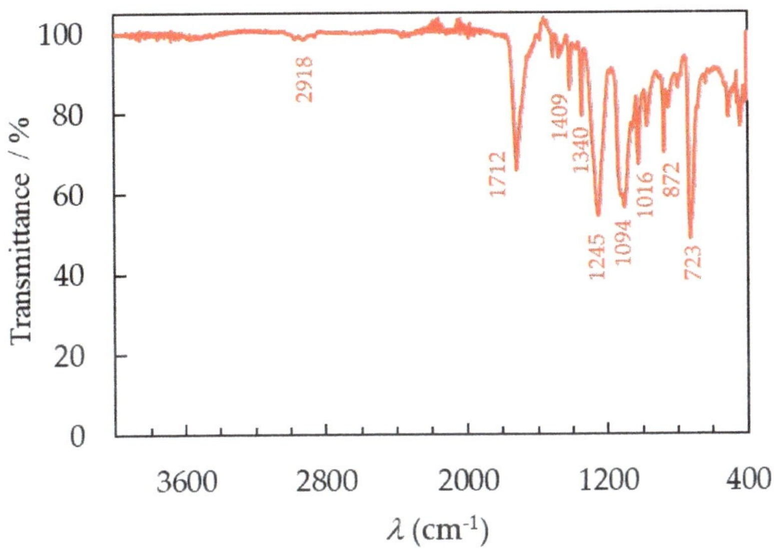

Figure 8. The FTIR spectra of recovered PET after separation of PV module.

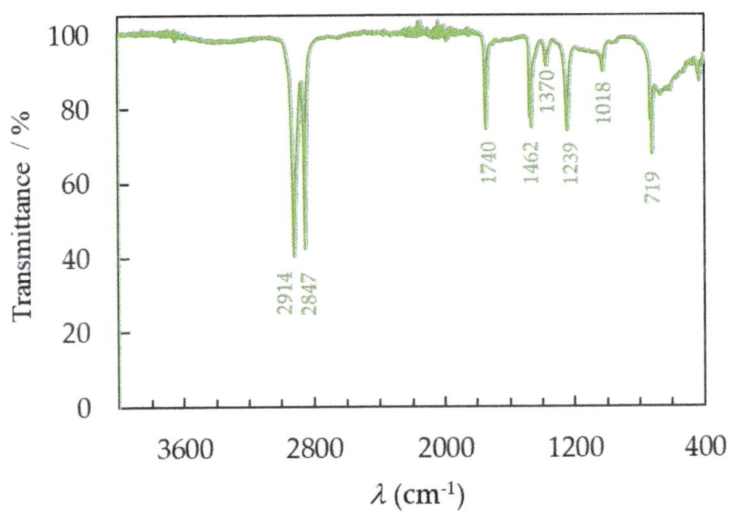

Figure 9. The FTIR spectra of recovered PVDF after separation of the PV module.

The laminated PV module also includes a white plastic layer that was placed between the EVA and PET layers. According to the literature, this can be assumed to be a thin PVDF layer. Nevertheless, the FTIR spectrum obtained from the sample, presented in Figure 8, does not validate this hypothesis. On the recorded spectrum, characteristic bands for the EVA polymer can be observed at 1740 cm^{-1} due to the bond stretching of the carbonyl group, C=O, which is attributed to the ester group. Two bands at 1239 cm^{-1} indicate the asymmetric C−O−C stretching vibrations in the ester group, and at 1018 cm^{-1}, they are associated with the stretching of the C−O bond. However, significantly more intense bands originating from the methylene $-CH_2$ group can be detected in the spectrum. The symmetric and antisymmetric stretching bonds of C-H are associated with two bands located at 2914 cm^{-1} and 2847 cm^{-1}, respectively. The presence of two bands at 1462 cm^{-1} and 1370 cm^{-1} can be attributed to the C−H bending. The rocking vibration of the $-CH_2$ group occurs at 719 cm^{-1}.

The FTIR spectrum indicates that the PVDF, which has been separated, contains leftover EVA polymer both on its surface and within its material. Neither mechanical nor chemical recycling methods achieved complete separation of the pure PVDF, compared to EVA and PET polymers. The PVDF spectrum does not fully match the spectrum seen in the literature [22] for the material extracted from the PV module. It has been verified that PVDF was obtained through recycling from the backsheet, but it is contaminated with EVA inclusions.

3.3.2. Elemental Analysis

The molar percentages of carbon (C), hydrogen (H), nitrogen (N), and sulfur (S) in the separated EVA, PET, and PVDF polymers are presented in Table 1. The polymers lack nitrogen and sulfur atoms in their structure, and these elements were not identified in the investigation.

Table 1. Elemental analysis of C and H for EVA, PET, and PVDF.

Polymer	Element (%)	Observed	Calculated
EVA	C	71.63	70.58
	H	10.76	10.59
PET	C	63.10	62.50
	H	4.33	4.17
PVDF	C	47.23	37.51
	H	5.31	3.15

The C and H contents of the EVA and PET polymers showed a satisfactory level of agreement with the composition derived from their structure. This provides additional proof that the polymers, which were separated using mechanical and chemical methods, were successfully obtained in their pure form. However, the quantified carbon and hydrogen composition of the PVDF layer exceeds the calculated values. This suggests that the layer has been contaminated with another polymer, EVA, that contains a high concentration of C and H atoms.

3.3.3. Differential Scanning Calorimetry

The results of the differential scanning calorimetry, DSC, measurements for EVA, PET and PVDF film samples recovered during the recycling of PV modules are shown in Figure 10. The measurements were performed over a wide range from −30 °C to 180 °C.

On the thermogram of the EVA, two overlapping endothermic peaks can be observed. The notable characteristic of these findings is the presence of two endothermic processes resulting from two groups of crystalline perfection. Published literature contains examples of this dual endothermic behavior [32–34]. The low-temperature endotherm corresponds to the melting of a group of imperfect, smaller crystallites, whereas the high-temperature endotherm is from the melting of larger, more regularly formed crystallites. The imperfect crystals are the result of the incorporation of branching and vinyl acetate comonomers into the polyethylene crystal lattice [33]. The peak temperature and total enthalpy of melting are summarized in Table 2.

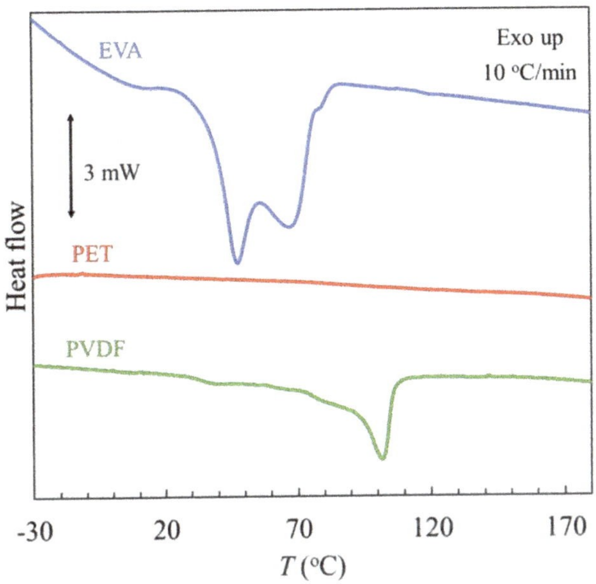

Figure 10. DSC analysis of separated polymers from the PV module.

Table 2. Thermal properties of separated polymers. Glass transition temperature, T_g; heat capacity at glass transition, $\Delta_g C_p$; melting temperature, T_m; and enthalpy of melting, $\Delta_m H$.

Polymer	T_g (°C)	$\Delta_g C_p$ ($J \cdot g^{-1} \cdot K^{-1}$)	T_m (°C)	$\Delta_m H$ ($J \cdot g^{-1}$)
EVA	-	-	(1) 47.4, (2) 66.9 (1) 49, (2) 72 [34]	55.0
PVDF	32.3	0.57	101.4	78.1

Standard uncertainties u are as follows: $u(T) = 0.3$ °C; $u(\Delta_g C_p) = 0.05$ $J \cdot g^{-1} \cdot K^{-1}$; $u(\Delta_m H) = 0.5$ $J \cdot g^{-1}$.

In the temperature range studied, no thermal transitions are observed on the thermogram of the PET, whether related to glass transition, melting, or crystallization of the plastic. This is consistent with the literature because for PET, a melting peak between 170 and 270 °C, with a maximum of 256 °C, is observable [35].

The thermogram of the PVDF material exhibits two distinct temperature transitions. At a temperature, $T_g = 32.3$ °C, the glass transition is observed. The melting peak is between 60 °C and 115 °C, with a maximum at $T_m = 101.4$ °C and a melting enthalpy value of 78.1 $J \cdot g^{-1}$. The recorded temperatures of both phase transformations are lower than for the literature values, where the glass transition between 55 and 65 °C and the melting peak in the range from 110 to 180 °C were observed [35]. Moreover, the measured melting enthalpy value is higher than the literature value, which is 30 J/g for a degree of crystallinity of about 30% [35]. The results suggest that the recovered material is not pure PVDF, but rather consists of unseparated impurities in the form of EVA, as confirmed by FTIR spectrum analysis.

All three investigated polymers do not undergo thermal decomposition up to the tested temperature of 180 °C in an atmosphere of the inert gas, nitrogen.

4. Conclusions

The present research develops an alternative method for the separation of PV modules once they have reached the end of their lifespan. First, the PV module was heated to the temperature $T = 170$ °C, which allowed for the softening of the EVA and the mechanical separation of the glass layer. The removed glass exposed the EVA encapsulation layer and improved contact with the solvent in the chemical step. The most dynamic effect of EVA swelling and PV module delamination was achieved using mechanical stirring and ultrasound. A very short delamination time, $t = 35$ min, during the chemical process was achieved under mild temperature conditions, $T = 35$ °C.

FTIR spectroscopy, elemental analysis, and DSC thermal measurements confirmed that the recovered polymers were EVA applied as a laminating layer and PET derived from the backsheet. The laminated PV module also included a white plastic layer that was placed between the EVA and PET layers. According to the literature, this can be assumed to be a thin PVDF layer. However, the conducted study did not confirm the acquisition of a pure PVDF polymer through the proposed separation method as it is mainly contaminated with EVA.

Hence, it is concluded from various characterization techniques that the separated EVA and PET polymers show quite similar properties as those of commercial polymers. Therefore, they can be reused for encapsulation or as a backsheet and in other applications in the packaging and textile industries.

Supplementary Materials: The following supporting information can be downloaded at: https://www.mdpi.com/article/10.3390/ma17040821/s1, Figure S1: PV module delamination (%) as a function of the density toluene solution; Figure S2: PV module delamination (%) as a function of the dynamic viscosity toluene solution; Table S1: Density, d, and dynamic viscosity, η, of toluene solution during delamination of PV module, measured at $T = 298.15$ K and $p = 100$ kPa.

Author Contributions: Conceptualization, M.K. and P.Ż.; methodology, M.K. and P.Ż.; validation, M.K., M.F., P.Ż. and M.M.; formal analysis, M.K. and M.F.; investigation, M.K., M.F., P.Ż. and M.M.; resources, M.K. and P.Ż.; writing—original draft preparation, M.K.; writing—review and editing, M.K., M.F., P.Ż. and M.M. All authors have read and agreed to the published version of the manuscript.

Funding: This research was supported by a grant from the Faculty of Automotive and Construction Machinery Engineering at the Warsaw University of Technology.

Institutional Review Board Statement: Not applicable.

Informed Consent Statement: Not applicable.

Data Availability Statement: The data presented in this study are available in the article and Supplementary Materials.

Conflicts of Interest: The authors declare no conflicts of interest. The funders had no role in the design of the study; in the collection, analyses, or interpretation of data; in the writing of the manuscript, or in the decision to publish the results.

References

1. Snapshot 2023—IEA-PVPS. Available online: https://iea-pvps.org/snapshot-reports/snapshot-2023/ (accessed on 5 December 2023).
2. Xu, Y.; Li, J.; Tan, Q.; Peters, A.L.; Yang, C. Global Status of Recycling Waste Solar Panels: A Review. *Waste Manag.* **2018**, *75*, 450–458. [CrossRef]
3. Heath, G.A.; Silverman, T.J.; Kempe, M.; Deceglie, M.; Ravikumar, D.; Remo, T.; Cui, H.; Sinha, P.; Libby, C.; Shaw, S.; et al. Research and Development Priorities for Silicon Photovoltaic Module Recycling to Support a Circular Economy. *Nat. Energy* **2020**, *5*, 502–510. [CrossRef]
4. Chowdhury, M.S.; Rahman, K.S.; Chowdhury, T.; Nuthammachot, N.; Techato, K.; Akhtaruzzaman, M.; Tiong, S.K.; Sopian, K.; Amin, N. An Overview of Solar Photovoltaic Panels' End-of-Life Material Recycling. *Energy Strategy Rev.* **2020**, *27*, 100431. [CrossRef]
5. Singh, S.; Powar, S.; Dhar, A. End of Life Management of Crystalline Silicon and Cadmium Telluride Photovoltaic Modules Utilising Life Cycle Assessment. *Resour. Conserv. Recycl.* **2023**, *197*, 107097. [CrossRef]
6. Cerchier, P.; Brunelli, K.; Pezzato, L.; Audoin, C.; Rakotoniaina, J.P.; Sessa, T.; Tammaro, M.; Sabia, G.; Attanasio, A.; Forte, C.; et al. Innovative Recycling of End of Life Silicon PV Panels: ReSiELP. *Detritus* **2021**, *16*, 41. [CrossRef]
7. Gahlot, R.; Mir, S.; Dhawan, N. Recycling of Discarded Photovoltaic Solar Modules for Metal Recovery: A Review and Outlook for the Future. *Energy Fuels* **2022**, *36*, 14554–14572. [CrossRef]

8. Song, B.P.; Zhang, M.Y.; Fan, Y.; Jiang, L.; Kang, J.; Gou, T.T.; Zhang, C.L.; Yang, N.; Zhang, G.J.; Zhou, X. Recycling Experimental Investigation on End of Life Photovoltaic Panels by Application of High Voltage Fragmentation. *Waste Manag.* **2020**, *101*, 180–187. [CrossRef] [PubMed]
9. Wang, R.; Song, E.; Zhang, C.; Zhuang, X.; Ma, E.; Bai, J.; Yuan, W.; Wang, J. Pyrolysis-Based Separation Mechanism for Waste Crystalline Silicon Photovoltaic Modules by a Two-Stage Heating Treatment. *RSC Adv.* **2019**, *9*, 18115–18123. [CrossRef] [PubMed]
10. Sasai, M.; Yamashita, T.; Inoue, D. Development of Low-Temperature Thermal Decomposition Recycling Technology from Photovoltaic Modules to Flat Glass Applications. *Jpn. J. Appl. Phys.* **2023**, *62*, SK1043. [CrossRef]
11. Fiandra, V.; Sannino, L.; Andreozzi, C.; Graditi, G. End-of-Life of Silicon PV Panels: A Sustainable Materials Recovery Process. *Waste Manag.* **2019**, *84*, 91–101. [CrossRef]
12. Fiandra, V.; Sannino, L.; Andreozzi, C.; Corcelli, F.; Graditi, G. Silicon Photovoltaic Modules at End-of-Life: Removal of Polymeric Layers and Separation of Materials. *Waste Manag.* **2019**, *87*, 97–107. [CrossRef]
13. Azeumo, M.F.; Conte, G.; Ippolito, N.M.; Medici, F.; Piga, L.; Santilli, S. Photovoltaic Module Recycling, a Physical and a Chemical Recovery Process. *Sol. Energy Mater. Sol. Cells* **2019**, *193*, 314–319. [CrossRef]
14. Subramanian, V.; Tembo, P.; Heninger, M. An Investigation of the Recovery of Silicon Photovoltaic Cells by Application of an Organic Solvent Method. *ECS J. Solid State Sci. Technol.* **2021**, *10*, 025001. [CrossRef]
15. Xu, X.; Lai, D.; Wang, G.; Wang, Y. Nondestructive Silicon Wafer Recovery by a Novel Method of Solvothermal Swelling Coupled with Thermal Decomposition. *Chem. Eng. J.* **2021**, *418*, 129457. [CrossRef]
16. Kang, S.; Yoo, S.; Lee, J.; Boo, B.; Ryu, H. Experimental Investigations for Recycling of Silicon and Glass from Waste Photovoltaic Modules. *Renew. Energy* **2012**, *47*, 152–159. [CrossRef]
17. Kim, Y.; Lee, J. Dissolution of Ethylene Vinyl Acetate in Crystalline Silicon PV Modules Using Ultrasonic Irradiation and Organic Solvent. *Sol. Energy Mater. Sol. Cells* **2012**, *98*, 317–322. [CrossRef]
18. Doi, T.; Tsuda, I.; Unagida, H.; Murata, A.; Sakuta, K.; Kurokawa, K. Experimental Study on PV Module Recycling with Organic Solvent Method. *Sol. Energy Mater. Sol. Cells* **2001**, *67*, 397–403. [CrossRef]
19. Brenes, G.H.; Riech, I.; Giácoman-Vallejos, G.; González-Sánchez, A.; Rejón, V. Chemical Method for Ethyl Vinyl Acetate Removal in Crystalline Silicon Photovoltaic Modules. *Sol. Energy* **2023**, *263*, 111778. [CrossRef]
20. Abdo, D.M.; Mangialardi, T.; Medici, F.; Piga, L. D-Limonene as a Promising Green Solvent for the Detachment of End-of-Life Photovoltaic Solar Panels under Sonication. *Processes* **2023**, *11*, 1848. [CrossRef]
21. Prasad, D.S.; Sanjana, B.; Kiran, D.S.; Srinivasa Kumar, P.P.; Ratheesh, R. Process Optimization Studies of Essential Parameters in the Organic Solvent Method for the Recycling of Waste Crystalline Silicon Photovoltaic Modules. *Sol. Energy Mater. Sol. Cells* **2022**, *245*, 111850. [CrossRef]
22. Dias, P.; Javimczik, S.; Benevit, M.; Veit, H. Recycling WEEE: Polymer Characterization and Pyrolysis Study for Waste of Crystalline Silicon Photovoltaic Modules. *Waste Manag.* **2017**, *60*, 716–722. [CrossRef] [PubMed]
23. Huang, W.H.; Shin, W.J.; Wang, L.; Sun, W.C.; Tao, M. Strategy and Technology to Recycle Wafer-Silicon Solar Modules. *Sol. Energy* **2017**, *144*, 22–31. [CrossRef]
24. Pagnanelli, F.; Moscardini, E.; Granata, G.; Abo Atia, T.; Altimari, P.; Havlik, T.; Toro, L. Physical and Chemical Treatment of End of Life Panels: An Integrated Automatic Approach Viable for Different Photovoltaic Technologies. *Waste Manag.* **2017**, *59*, 422–431. [CrossRef] [PubMed]
25. Spinella, L.; Bosco, N. FTIR Investigation of EVA Chemical Bonding Environment and Its Impact on Debond Energy. *IEEE J. Photovolt.* **2019**, *9*, 790–795. [CrossRef]
26. Klemchuk, P.; Ezrin, M.; Lavigne, G.; Holley, W.; Galica, J.; Agro, S. Investigation of the Degradation and Stabilization of EVA-Based Encapsulant in Field-Aged Solar Energy Modules. *Polym. Degrad. Stab.* **1997**, *55*, 347–365. [CrossRef]
27. Coulter, D.R.; Cuddihy, E.F.; Plueddeman, E.P. *Chemical Bonding Technology for Terrestrial Photovoltaic Modules*; NASA: Washington, DC, USA, 1983.
28. Feng, Y.; He, Y.; Zhang, G.; Wang, S.; Wei, N.; Zhang, T. A Promising Method for the Liberation and Separation of Solar Cells from Damaged Crystalline Silicon Photovoltaic Modules. *Sol. Energy Mater. Sol. Cells* **2023**, *262*, 112553. [CrossRef]
29. Huang, Q.; Yuan, W.; Guo, Y.; Ke, Q. Thermal Separation of Plastic Components from Waste Crystalline Silicon Solar Cells: Thermogravimetric Characteristics and Thermokinetics. *J. Air Waste Manag. Assoc.* **2023**, *73*, 853–864. [CrossRef] [PubMed]
30. Chitra; Sah, D.; Lodhi, K.; Kant, C.; Saini, P.; Kumar, S. Structural Composition and Thermal Stability of Extracted EVA from Silicon Solar Modules Waste. *Sol. Energy* **2020**, *211*, 74–81. [CrossRef]
31. Han, H.; Yan, H.; Wang, X.; Zhang, K.; Huang, J.; Sun, Y.; Liu, J.; Verlinden, P.J.; Altermatt, P.; Liang, Z.; et al. Analysis of the Degradation of Encapsulant Materials Used in Photovoltaic Modules Exposed to Different Climates in China. *Sol. Energy* **2019**, *194*, 177–188. [CrossRef]
32. Tsocheva, D.; Tsanov, T.; Terlemezyan, L. Structure of Composite Films Containing Polyaniline Studied by DSC. *J. Therm. Anal. Calorim.* **2001**, *66*, 415–422. [CrossRef]
33. Agroui, K.; Maallemi, A.; Boumaour, M.; Collins, G.; Salama, M. Thermal Stability of Slow and Fast Cure EVA Encapsulant Material for Photovoltaic Module Manufacturing Process. *Sol. Energy Mater. Sol. Cells* **2006**, *90*, 2509–2514. [CrossRef]
34. Marcilla, A.; Reyes-Labarta, J.A.; Sempere, F.J. DSC Kinetic Study of the Transitions Involved in the Thermal Treatment of Polymers. Methodological Considerations. *Polymer* **2001**, *42*, 5343–5350. [CrossRef]
35. Oreski, G.; Wallner, G.M. Aging Mechanisms of Polymeric Films for PV Encapsulation. *Sol. Energy* **2005**, *79*, 612–617. [CrossRef]

MDPI AG
Grosspeteranlage 5
4052 Basel
Switzerland
Tel.: +41 61 683 77 34

Materials Editorial Office
E-mail: materials@mdpi.com
www.mdpi.com/journal/materials

www.ingramcontent.com/pod-product-compliance
Lightning Source LLC
Chambersburg PA
CBHW061630210926
51850CB00009B/2495
* 9 7 8 3 7 2 5 8 5 9 9 7 9 *